Understanding
Weatherfax

Understanding
Weatherfax

Mike Harris

Second edition

SHERIDAN HOUSE

Contents

This edition published 2005 by
Sheridan House Inc.
145 Palisade Street
Dobbs Ferry, NY 10522
www.sheridanhouse.com

Copyright © 2005 by Mike Harris

Library of Congress Cataloging-in-Publication Data

Harris, Mike
  Understanding weatherfax / Mike Harris.
    p. cm.
  Includes index.
  ISBN 1-57409-215-4
  1. Weather forecasting-Equipment and supplies.
  2. Weather forecasting-Charts, diagrams, etc.
  3. Fax machines. 4. Facsimile transmission-Equipment and supplies. I. Title.

QC995.48.H37 2005
551.63'028'4-dc22

2005010856

ISBN   1-57409-215-4

Printed in Great Britain

Acknowledgements

'Weather forecasting is a sharing of ideas' says Bob McDavitt, my friend and mentor from
New Zealand MetService. He refers to the way that forecasters draw on a wealth of diverse
sources from computer models, to ships at sea, to satellites in space. Writing this book
has also been a sharing of ideas, and I'm most grateful to the many people who provided
contributions. Unfortunately, this is too many to mention individually, but particular thanks
are due to:

Bob McDavitt of New Zealand MetService.
Bill Hall (G4FRN), net controller of the United Kingdom Maritime Net.
Paul Hayden, for help with satellite reception techniques.

Last but by no means least, a special thanks to Di, my wife, for her painstaking preparation
of the illustrations.

Glossary of Terms and Abbreviations

AGC   Automatic gain control.
Anabatic wind   Local wind blowing up slopes heated by sunshine.
ANAL   Analysis.
Anticyclone   A high pressure system.
 Weak high: central pressure less than 1015 hPa
 Strong or intense high greater than 1030 hPa.
APT   Automatic picture transmission.
AVHRR   Advanced high resolution radiometer.
CAT   Clear air turbulence.
CB   Cumulonimbus clouds.
CTH   Cloud top heights.
Cyclonic system   A low pressure system (*see Depression*).
 Shallow low: central pressure greater than 1000 hPa;
 Moderate low: central pressure less than 1000 hPa but greater then 980 hPa;
 Deep or intense low: central pressure less than 980 hPa.
Depression   A low pressure system usually associated with frontal activity.
Deviation frequency   Half the difference between the frequency for white and the frequency for black.
Dew point   The temperature at which an air sample containing water vapour is saturated. Above the dew point the air is able to absorb more water. Below it, water condenses out as fog or water droplets.
Doldrums   See ITCZ.
ECMWF   A forecast model created by the European Centre for medium range weather forecasts.
FL XXX   Flight level – with height in hundreds of feet, eg FL 100.
FREQ   Frequent.
FREQ EMBD CB   Frequent cumulonimbus clouds with little or no separation which are embedded in an area of other layered cloud.
Fronts   Meeting of cold and warm air, bringing disturbed weather.
FSK   Frequency shift keying.
GASP   A forecast model: Global Assimilation and Spectral Prognosis.
Geostrophic wind   Theoretical wind strength estimated from isobar spacing only (excluding friction, latitude, isobar curvature and local effects).
GMT   Greenwich Mean Time – see UTC.
GRIB files   Gridded Binary Data files. Compact files commonly used for

chart overlays showing wind arrows, sea temperatures etc.

GOES   Geostationary operational environmental satellites.

hecto Pascal (hPa)   Unit of pressure equivalent to 1 Millibar.
  1 hPa = 100 Pa.

HF   High frequency (3 to 30 MHz); synonymous with SW.

Horse latitudes   Latitudes in the range 30 to 40 degrees where sub-tropical highs may give long periods of light, variable or calm winds.

HRPT   High-resolution picture transmission.

Hurricane   Local term for a cyclone.

IAC Fleet Codes   A system of transmitting a synoptic chart using plain text 5-figure groups.

IHO   International Hydrographic Office.

IMO   International Maritime Organisation.

IOC   Index of co-operation.

IR   Infra-red.

Isobar   A line of equal barometric pressure.

ISOL EMBD   Isolated individual cumulonimbus clouds which are embedded in an area of other layered cloud.

ISOLD   Isolated.

Isotach   A line of equal wind velocity; shown on some upper level air charts.

ITCZ   Inter tropical convergence zone (see pages 30–1).

JMA   Japan Meteorological Agency forecast model.

Katabatic wind   Wind cascading from mountainous regions. Caused by air becoming heavier after losing heat by radiation.

LAPS   Local Area Prediction System forecast model.

LEOS   Low Earth Orbit Satellite.

LRIT   Low-rate information transmission.

LRPT   Low-resolution picture transmission.

MET AREA   Meteorological area (as defined by WMO) and identical to NAV AREA.

Millibar   Unit of pressure equivalent to 1 hecto Pascal (hPa).

MSL   Mean sea level.

NAV AREA   Navigational area (defined by IMO and IHO).

NEPHANAL   Cloud cover analysis chart.

NMRS   Numerous.

NOAA   National Oceanographic and Atmospheric Administration.

NOGAPS   Navy Operational Global Atmospheric Prediction System.

NORAD   North American Air Defence Command.

NORAPS   Navy Operational Regional Atmospheric Prediction System.

OCNL   Occasional.

OCNL EMBD CB   Occasional well-separated cumulonimbus clouds which are embedded in an area of other layered cloud.

Pascal (Pa)   Unit of pressure; I Pa = I Newton/sq metre.
PLL   Phase locked loop
POES   Polar Orbiting Environmental Satellites.
PROG   Prognosis.
RAFC   Regional Area Forecast Centre.
Relative humidity   Ratio of specific humidity to its saturation value at ambient temperature.
RF   Radio frequency.
SCT   Scattered.
SFC   Surface.
Shear line   Zone across which wind direction changes sharply.
SIG WX   Significant weather features.
Specific humidity   Concentration of water vapour in air by weight.
STNRY   Stationary.
SW   Short wave (10 to 100 metres) synonymous with HF.
TAF   Terminal aviation forecast.
Terminator   On a satellite image, the twilight division between sunlight and darkness.
TIROS   Television and Infra-red Observational Satellite.
TNC   Terminal node controller.
UKMO   United Kingdom Met Office forecast model.
UTC   Universal Co-ordinated Time; for the purpose of this book it is synonymous with GMT or Zulu time.
VHF   Very high frequency.
VIS   Visual.
WEFAX   Weather facsimile.
WX   Weather.
Zulu time   Synonymous with UTC.

Preface

A basic computer and radio receiver are the only tools needed to gain access to the enormous amount of weather information continually broadcast by the world's short wave fax broadcasters. In addition to the familiar synoptic charts showing isobars, fronts, high and low pressure weather systems, there are charts of wave heights, sea temperatures, high altitude wind conditions, satellite pictures and almost any meteorological detail one could wish for.

Suppliers of weatherfax software seldom give much information on how to interpret charts, and books on meteorology are often all-encompassing and devote little to the subject. In writing this book I have attempted to fill the gap. It is not an 'all you need to know about the weather' type of book, but instead the aim has been to provide only the essentials needed to set up a weatherfax receiver and derive useful forecast information from the results. Its underlying objective is to help promote the habit of seeking relationships between chart details and the actual weather you experience – an important part of understanding weatherfax.

Chapter 1

Weatherfax and Satellite Image Technology

To most of us, a weather chart is what you see on the television or in newspapers, but how do you get a current chart if you are at sea or at some remote location? The solution lies in using either a telephone or short wave radio-based image receiver. Of the two, telephone fax is the most familiar because of its widespread use in offices and homes.

This, though, is not the only method of sending images over phone lines. With the huge expansion of the Internet many people regularly use computers to send quite detailed colour pictures at relatively low cost.

Radio fax, on the other hand, is a much older technology, having been developed many decades ago. However, it has several distinct advantages and remains the system of choice for most marine users. Obtaining good-quality pictures and using the system effectively requires some understanding of how it works, and thus the latter parts of this chapter explain how short wave radio fax charts are transmitted and the types of equipment needed to receive them.

Telephone fax (telefax)

For many years, meteorological offices have provided a telephone service where subscribers can listen to the latest forecast for a particular area. In recent times some have added fax facilities where, in response to a recorded message, subscribers use the key pad to select their choice of fax chart. In parts of Europe, Australia and New Zealand the service is widely available and known as *MetFax* or *AirFax*. Advantages of the service are that it's easy to set up and operate and it provides clear quality images. Telefax charts are frequently updated and are often large scale giving good coverage of local details and are ideal for coastal cruising.

Internet sources of weather charts

Most National Meterological services have established their own World Wide Web pages and provide a wealth of detailed weather information

1

to anyone able to access the Internet. At present, much of the freely available information is intended for general public interest though if you are willing to explore a little there are also plenty of synoptic charts and satellite pictures. Image quality can be good but watch out for sites set up for demonstration purposes only, showing material that is spectacular in appearance but out of date. As with many parts of the Internet the main difficulty with weather information is the sheer number of sources and the frequency with which they change. It would be impossible to give a list that would not be outdated in a few weeks though some sites eg 'Weather Lists' are specifically arranged to provide connections to other sources.

The USA and Canada are particularly well served by Internet weather sources. If you are browsing for the first time, a search under 'US Coast Guard' or 'Weather Lists' is sure to bring results. The US Coast Guard telecommunications page in particular includes links to other useful sites as well as links to High Seas radio stations. These provide listings of their current frequencies and broadcast schedules and those with a radio fax service may offer downloadable graphic files of their most recent charts.

An obvious difficulty with any method of obtaining weather charts by phone line is that you need a phone, which can present difficulties if you are at a remote land site or at sea. Many commerical marine radio stations operate a telephone link call system on VHF and HF. In principle, this could be used to receive faxes anywhere in the world but one of the difficulties is that marine radios aboard most smaller vessels are simplex sets, whereas telephone fax signals are fully duplex[1]. In an effort to overcome this problem, a few radio stations have installed equipment enabling them to transmit phone fax in a modified format but the service is not widely available and needs special decoding equipment. Satellite phones are another possibility and again, in principle, any service available to land line subscribers could also be provided to mobile and marine stations. In practice, this is only attractive if circumstances justify the very high equipment and call costs and the comparatively low data transfer rate.

No doubt as technology evolves, the next few years will see significant improvements in price and data speed but for the immediate future, cellular phones also have a useful role to play. Compared to land line phones, cellular charges are still high but well below the cost of satellite services. Data transfer rates are quite modest and the service is only available in the more developed countries and extends little more than 10 or 20 miles from the coast. None the less, the service has good potential for coastal

---

[1]Duplex operation occurs between two radio stations that are able to transmit and receive at the same time. Simplex operation is used on radio equipment that cannot simultaneously transmit and receive.

traffic and, in principle, with international billing arrangements your service and account in one country could be picked up in another when you arrive within range.

Weather charts from short wave radio broadcasts

The method of broadcasting weatherfax images over short wave (HF) radio frequencies is totally unlike that for telephone fax. Essentially the differences are:

- Each chart takes 8 to 20 times longer to transmit.
- The radio broadcasts can be picked up by an unlimited number of listeners.
- The service is generally provided to improve safety at sea and has no user charges.
- A simplex radio link is used so a 'receive only' radio is all that's needed.

Before discussing the types of equipment used to receive HF fax images, it is useful to know a little of the history of the method and something of the make-up of fax transmissions.

Drum scanning

Much of the terminology attached to current radio fax practice has been in use for many years and was originated at a time when equipment was very different from that in use today. In a modern context, the terms can appear somewhat obscure, so it is helpful to understand a little of the design of early equipment which, being mechanical, is often easier to appreciate than the hidden workings of an electronic assembly.

On early weatherfax transmitters, the chart was attached to the outside of a cylindrical drum measuring 152 mm in diameter and 594 mm in length. These dimensions are only important in so far as they relate to the proportions of present-day charts which, incidentally, are rather different from the computer screens on which they are often displayed. With the chart's length running parallel to the axis, the drum was rotated at a constant speed, usually 120 revolutions per minute. A light sensor was arranged to pass slowly along the drum axis and hence the length of the chart. With each rotation it scanned a single narrow line along the width of the chart. Its output was used to produce an audio tone of either 1.5 kHz or 2.3 kHz, depending on whether it was passing over a black or white part of the chart. Instead of a drum, modern fax transmitters use a flat-bed scanner, though the term *drum speed* still persists.

Most weatherfax stations use a drum speed of 120 rpm, though there are other standards in existence. To obtain an image that looks correct, it is important that fax-receiving equipment is set to receive at the same speed.

3

Make-up of a fax signal

Fax signals on short and long wave frequencies are normally transmitted as a frequency shift keyed signal (FSK). On simple charts that only include black lines on a white background, the transmitter switches between a pair of frequencies spaced 800 Hz apart (300 Hz apart on long wave). The upper of the pair corresponds to the white part of the chart, while the lower represents the black. (A very few fax stations use the lower part for white and the upper one for black.) A few stations also transmit pictures, usually originating from satellite images of clouds and temperatures. In these cases, the various shades of grey are represented by frequencies between the black and white extremes.

To the casual listener tuning across the band, a fax signal sounds like a repeated scratching sound – much like the repeated rubbing of a woollen glove over a rough brick or the tearing apart of a velcro strip. The noise is of the rapid switching between black and white tones, and exactly how it sounds depends on the type of picture being transmitted.

Typically, a weather chart will take between 8 and 20 minutes to send and, in addition to the body of the image just described, includes up to 5 additional, audibly distinguishable parts, as follows:

I Start Tone   This is a tone formed by fast, regular switching

Another parameter affecting the appearance of the received image is the index of co-operation (IOC). This refers to the scan line density and, if set incorrectly, results in an image that is vertically stretched or compressed, though still readable. Most weatherfax stations use an IOC of 576.

Receiving equipment

For weatherfax, the choice is essentially between a dedicated, self-contained weatherfax receiver or the combination of a good-quality radio with a computer interface and weatherfax software. The dedicated fax receiver has the advantage that the whole installation is included in one box. Operation is made as simple as possible and running the unit does not tie up other pieces of equipment. A common feature, which is more trouble to arrange with a separate radio and computer, is the ability to remain idle in a standby mode drawing minimal current, and then to switch on automatically at a pre-programmed time and frequency. This is between the black and white frequencies and it is used to start certain types of automatic receivers. The switching frequency, usually 300 Hz for most weatherfax transmissions, is used to indicate the index of co-operation (see page 18).

2 Phasing Signal or Sync Pulse   This follows briefly after the start tone and is heard as a repeated blip. On screen it appears as a rectangular block that marks the beginning of a scan line. Ideally, this should be positioned at the extreme left-hand, top corner of the screen. If it appears in the middle, the received image will be split into two parts; this normally causes no difficulties as most software contains a facility for rejoining (rejustifying) the two halves (see Fig 1.4).

3 Test Scale or Tone Bar   Not all stations transmit a tone bar but, when present, it is heard as a repeated whistle or steadily changing frequency. On a correctly tuned screen it appears as a narrow band of grey progressing from black to white.

4 Image Body   The image itself, the longest part of the transmission, which lasts between 8 and 15 minutes.

5 Stop Tone   This sounds rather like the start tone, but is sent at 450 Hz and tells the receiving equipment that the fax is complete.

6 Unmodulated Carrier   Some stations carry on transmitting a continuous black tone signal for a few seconds or minutes after the stop tone.

useful if the faxes you need are broadcast at inconvenient times.

Within the last few years, the use of computers for receiving weatherfax has increased enormously. If you already have a good communications receiver and a computer, the ability to receive weatherfax can be added for just the cost of the software and interface. It's convenient to be able to print received charts for easy reference, though a printer is not an essential part of the set-up as they can be saved on disk and displayed on screen when required, or even incorporated within other documents or software. When compared to a dedicated fax receiver, a separate radio/computer installation will require slightly more skill to set up, and results will be largely dependent upon the quality of the receiver.

Radios for fax
On the whole, any good-quality amateur or marine band transceiver or communications receiver is likely to make a good fax receiver. A few of

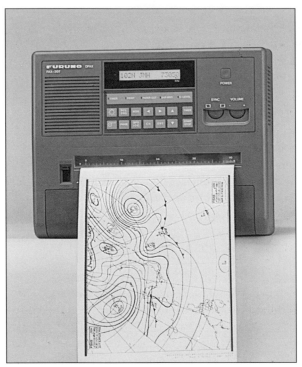

Furuno weatherfax receiver (By courtesy of Furuno).

the better-quality radios intended for broadcast listening can also produce acceptable results, but many of the cheaper sets in this category are simply too noisy or drift in frequency too much to be worth considering. Important features to look for are:

- Ability to resolve single side band signals.
- Coverage of the short wave bands 0.5 to 25 MHz.
- Fine tuning in increments of not more than 100 Hz.
- Good frequency stability.
- Good sensitivity and selectivity.
- Low internal noise level.

Analogue radios, unless of particularly good quality, are inclined to drift off frequency but have the advantage that internal noise levels are low. On the other hand, digitally synthesised receivers, using phase locked loop (PLL) tuning, usually have excellent frequency stability but, in cheaper sets, are often so noisy that all but the strongest signals are swamped. As with so many things, you get what you pay for.

Grounding

Grounding is not as critical for receivers as it is for transceivers. None the less, a good radio frequency ground can make an enormous difference to the received signal. For land-based stations the solution could be to connect the receiver ground terminal to a buried metal sheet. It should be as large as practicable and preferably well submerged beneath damp ground. At sea, a connection to a steel hull or the armature of a concrete boat is also effective, but with wooden or glass fibre hulls possible alternatives are connections to:

- An encapsulated metal keel.
- Use of a large area of metal foil or mesh against the inside of the hull or laid up beneath the last layer of glass fibre.
- A square metre or more of copper sheet secured to the outside of the hull.
- A sintered metal earthing plate bolted to the outside of the hull.

The computer and installation

The computer specification depends entirely on the requirements of the software you have available. Software is available for most computers including the Apple Mac, but far more has been written for IBM compatible than other types. Even some of the earliest personal computers, based around the 8086 processors, with a single serial port, one floppy disk drive, 520K RAM, no hard drive and early monochrome screen graphics, can still be used. Early liquid crystal and gas plasma screens were hardly good enough but, to some extent their short comings can be overcome by installing a compatible printer to display the pictures.

A critical aspect of the computer that is not always mentioned in the sales literature concerns the amount of electrical noise it generates. Interference produced from within the computer may travel to the radio through power or interface leads or it can be radiated through space. Once in the radio, it is amplified along with the other signals and is passed back to the computer, where it appears as a textured pattern overlaying the received picture, possibly obscuring it completely. With improved manufacturing standards, this type of interference is not the problem that it was on earlier computers.

The type of interference shown in Fig 1.2 (page 10) could also have originated from some other piece of electrical equipment but, in either case, good installation practice can go a long way towards eliminating these difficulties. In particular:

- Do not run power lines close to antennas or signal lines between the radio and computer.

The antenna

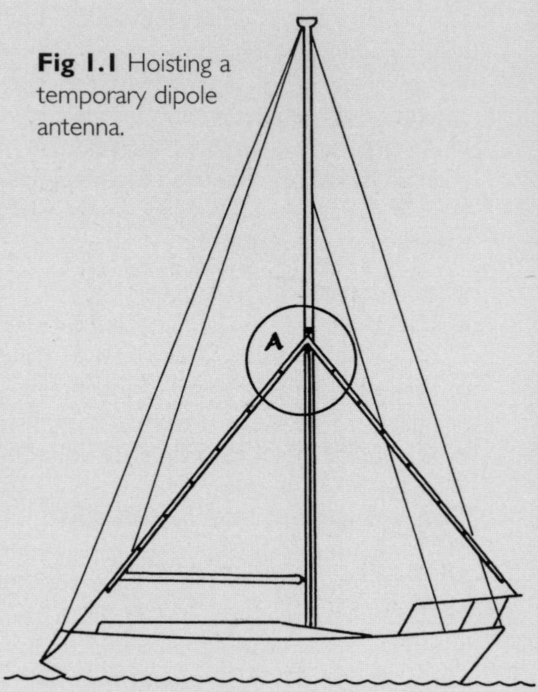

Fig 1.1 Hoisting a temporary dipole antenna.

If your radio is part of a communications transceiver, then beyond making sure it is tuned correctly, in all probability there will be little you can do to improve it for fax. It hardly bears mentioning, but when tuning for a fax frequency, do consider others that might also be listening and use a dummy load or some other method that avoids transmitting on the frequency.

Good transmitting antennas usually perform well as receiving antennas, though the converse is not always true. Quite often it seems that you can string up almost any odd length of wire and it brings in good signals. If it also gives you a fax quality that you are happy with, then all well and good. However, if you are using a good radio but the signals are still weak, swamped by noise or by other stations, some attention to the antenna design could be the answer.

Firstly, most man-made electrical noise (eg interference from motors etc) is vertically polarised, and will be less of a problem if the receiving antenna can be arranged horizontally. On a small boat this may not be possible and the use of an insulated section of a mast stay, a vertical whip or temporary wire hoisted from a halyard may be the only possibilities.

Fig 1.1 shows design details of a simple dipole, so named because it consists of two limbs fed at the centre. These work best when cut to a length that relates to the frequency of signals you intend to receive. The length (L) of each leg in metres can be calculated from the following formula:

$$L = 95\% \text{ of } (300/\text{Frequency in MHz})/4$$

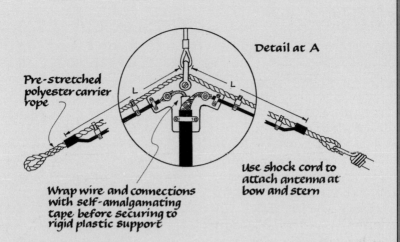

Detail at A

Pre-stretched polyester carrier rope

L   L

Wrap wire and connections with self-amalgamating tape before securing to rigid plastic support

Use shock cord to attach antenna at bow and stern

Active antennas

If you do not have space to accommodate a full-size antenna – perhaps because you need to receive long wave length signals – a short length of wire and an active antenna may be the solution. Active antennas are not really antennas at all, but wide band amplifiers that effectively increase the strength of all received frequencies. Their power requirements are usually so small that they can run for several months on a single set of batteries. Often they are enclosed in a small weatherproof case and mounted close to the antenna base.

Active antennas are not always the panacea for bad reception that their advertisers might have us believe. The main problems stem from the fact that they amplify *all* signals alike: the strong, the weak, the wanted and the unwanted. Thus background noise and interference are also increased, and if a strong signal is present its amplified strength could overload the receiver.

To some extent the solution lies in providing additional tuning and filtering in the form of an antenna-tuning unit, similar to those fitted to transceivers. However, for receive-only use, it would not need to handle high power levels and could be relatively small and inexpensive.

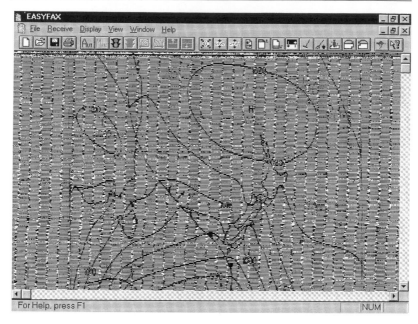

Fig 1.2 Fax picture severely affected by interference. The texture pattern is characteristic of machine noise.

- Use screened cables for all interconnections and make sure that one end of the screen is connected to an effective ground.
- Thread ferrite beads along power leads at the point of entry to the computer.
- If problems persist, try relocating the computer. If possible, on the other side of a steel bulkhead. If not, a few metres extra separation can make a large improvement.

The radio/computer interface

With early weatherfax software, the computer's serial port (otherwise known as a COM or RS232 port) provided the entry point for audio signals from the radio. Since this is primarily designed to connect digital peripherals such as printers, modems, scanners etc, its use for analogue audio signals involved the use of special software techniques and an interface with a small amount of electronics. Under the Windows® operating system this became quite complex but, fortunately, a better alternative presented itself when sound cards became a standard accessory for all computers.

A sound card's microphone input is specifically designed for feeding radio audio signals into a computer and is an ideal input point for audio fax signals. At the time of writing, there are several inexpensive fax programs (see Appendix 8) that make use of it, and all that's required is a coaxial cable connection. Nothing could be simpler.

Before leaving this section, mention should be made of Multimode Data Decoders (or TNC) such as the Kantronics KAM, SCS PTC-II, AEA PK-232, or MFJ 1276 or 1278B. These small but highly versatile units are dedicated to converting a range of types of radio data and so have a purpose-built audio input port. Decoded data is fed into the computer's serial port for display on the screen. In addition to weatherfax, they are capable of receiving and transmitting a range of data types such as AMTOR, PACTOR, G-TOR, PACKET and Morse. The penalty for such versatility is increased complexity, though for anyone needing to use HF e-mail, they're essential.

Software

There are many programs available for the reception of fax and satellite images (see Appendix 8). Clearly, a first requirement is to choose one that's written to run under the operating system used by your particular computer but, further than this, it needs also to be matched to your particular hardware. There is no point in choosing a program written for 32 bit Windows® (eg 95, 98, NT, ME, or 2000) if your computer is an early PC and has sufficient RAM to run only Windows® 3.1. If you have a sound card, then a choice of software that makes use of the card could save trouble with hardware.

Fax and satellite image reception software is available through a variety of sources including retail outlets, mail order, magazines, or over the Internet. Expect to pay more for a product that's distributed through a big name distributor and comes boxed with a printed manual and every wire, plug, and socket that you might need. Price, however, often has little bearing on quality, as some very good software products are available through the smaller Internet sites.

When buying from any source, take care to ensure that in-depth technical support is readily available should you need it. A good question to ask of the sales person is if they have actual hands-on experience of the product themselves and do they use it on their own computer? With Internet buying, the physical location of a firm is of little importance and it doesn't matter if they don't have a branch in your home country. Far more critical is that they are able to provide effective on-line support, so before buying check that their website shows how to contact their support section. Make sure you've carefully read the product details, particularly any FAQs (Frequently Asked Questions), and then test the system by e-mailing a question of your own. No marks for firms that send

11

only automated replies, and if they're worth their salt, expect a considered response within a couple of days.

When deciding on which software to buy, it's a good idea to see it running before committing yourself. A few of the better retailers have showroom demonstrations, and over the Internet some sites offer free trial versions. Some features to look for are:

- Ability to save faxes to disk or printer.
- Reversal of background and foreground colours.
- Progressive 'cleaning' of interference from received images.
- Magnification of selected parts of an image.
- 90° rotation of received images.
- Ability to trim tops and bottoms of images.
- Reattachment (justification) of vertically split images.
- Automatic starting and stopping or received picture signals.
- Pre-programming of scheduled start times.
- Grey scale resolution.

Receiving fax signals

The majority of short wave weatherfax stations are located on high frequency (HF) bands between 4 MHz and 20 MHz. The weatherfax station browser in Appendix 1 gives a list of the more popular stations, together with their operating frequencies. For more comprehensive details, consult one of the reference books in Appendix 9.

If after tuning your radio to a fax station you hear nothing, does it mean

Frequency band (MHz)	Daytime range (miles)	Night-time range (miles)
1.5 to 3	300	1000
3 to 6	400	1500
6 to 10	600	2000
10 to 16	> 200, < 1800	Unlimited in the direction of the sun
16 to 23	> 300, < 3000	Unlimited in the direction of the sun
22 to 30	Unlimited but depends on sun spot activity	Poor after sunset

Table 1.1 Approximate ranges obtainable on short wave frequencies.

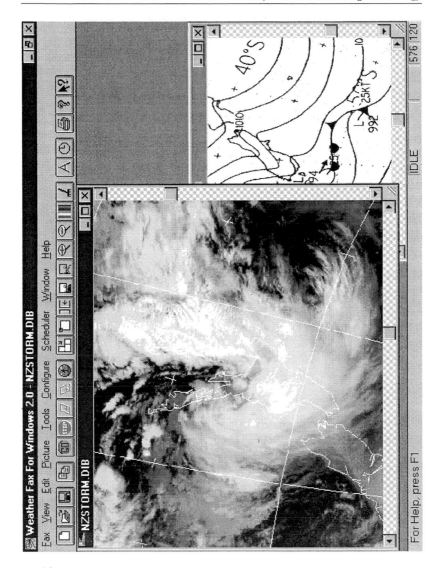

Fig 1.3 Weatherfax for Windows℗ showing a satellite image of a New Zealand storm together with a conventional fax image.

they are off the air or the schedule list is wrong? Not necessarily. The first possibility to consider is that perhaps you caught the station during a pause between transmissions. Often there are gaps of 10 or more minutes between faxes and some stations only operate on a part-time basis. Occasionally, the time is used to transmit a carrier wave, so you may hear only a continuous whistle, but other stations just switch off.

Next, there are the characteristics of short wave propagation to consider, which are very dependent upon the concentration of ionised gases in the upper atmosphere. This, in turn, is determined largely by the sun – some frequencies becoming more effective at night, or are best only at peaks in the 11-year sunspot activity cycle. These effects are roughly summarised in Table 1.1 but more reliable forecasts can be obtained by using a Propagation Prediction Program or from one of the shortwave magazines that publish prediction data.

Tuning the radio for fax

Tuning into fax signals is easy enough when you know what to listen for, but be prepared for a little trial and error if good results are not immediately forthcoming. Firstly, fax signals are best received in a side band mode. Usually, upper side band is used, though lower side band works just as well except that in one case the picture will appear as black on a white background and as white on a black background in the other. Most software includes a facility for reversing the colours so they can be switched later if you prefer.

Another point to bear in mind is that the actual frequency you set on the receiver dial is likely to be slightly different from that given in the official frequency lists. On upper side band it will be about 1.9 kHz below the published frequency and a similar amount above on lower side band. In practice it is better to fine tune the radio (or if possible the software) to get the best picture quality rather than aiming for a spot frequency.

Most communications quality receivers have a variety of controls for reducing background noise, fading, etc. These are primarily intended to improve the quality of received speech. With fax signals, such controls need to be used with caution:

- If you have tone controls, turn the treble well up and reduce the bass to give the widest audio band width.
- Turn off any noise blanking. Fax signals can sound like noise to a noise blanker circuit, which may remove part of the signal.
- Switch off the automatic gain control (AGC).
- Fax signals contain frequencies between 1200 and 2400 Hz. Use narrow band and variable width filters with care to prevent clipping upper and lower limits.
- Watch out for notch filters that remove fixed frequency whistles as they can also wipe out a particular tone from received pictures.

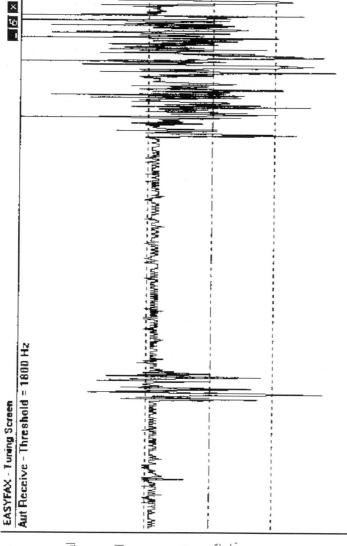

EASYFAX - Tuning Screen

Aut Receive - Threshold = 1800 Hz

Fig 1.4 Tuning display. To help with tuning, most software includes an oscilloscope type display giving a graphical representation of the incoming signal. The three broken horizontal lines represent the mid point and upper and lower transitions. All signals appearing above the top line represent black parts of the image and all below the lower, white. Fine tuning the radio shifts the wave form to move up or down the screen though a similar affect can be achieved with software controls that move the threshold lines.

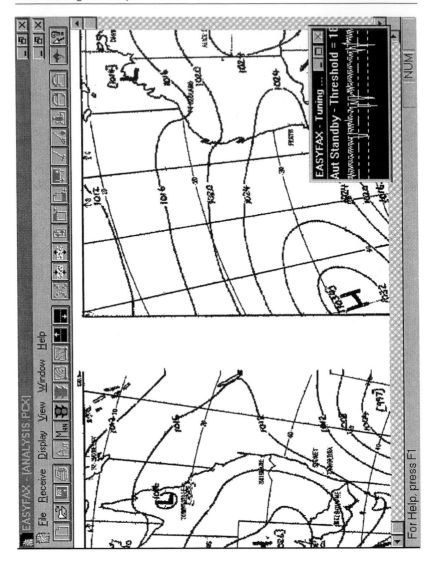

Fig 1.5 Incorrect synchronisation.

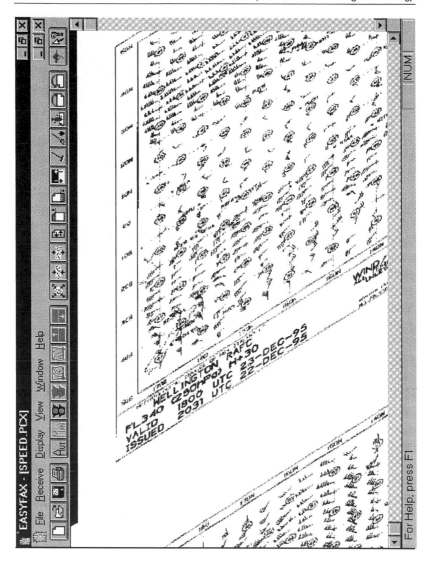

Fig 1.6 Incorrect receive speed setting.

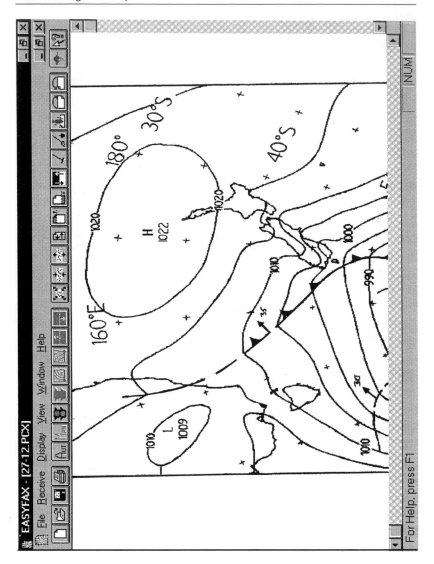

Fig 1.7a Correct index of co-operation.

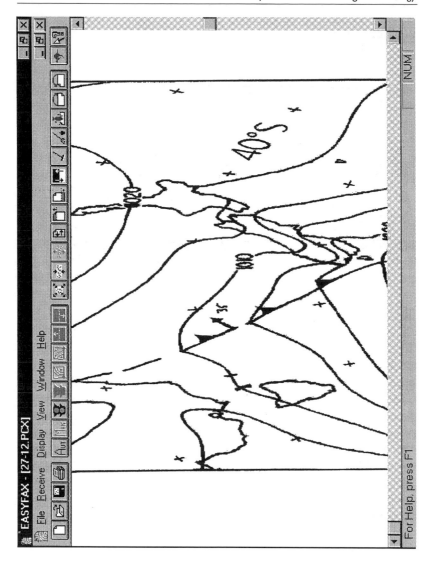

Fig 1.7b Incorrect index of co-operation. The picture appears stretched.

Tape recording fax broadcasts

If you have a tape recorder and timer that can be used to record broadcasts, it would be useful to be able to record inconveniently timed faxes to be displayed later. Unfortunately the idea is unlikely to be successful as the speed of most tape recorders is too variable to produce a stable image. If your computer has a sound card it is possible to record broadcasts as a sound file. Saved in this way they can produce good-quality pictures, though the amount of memory consumed can be prohibitive.

Configuring the software

As was mentioned earlier, weatherfax software design varies enormously, though every program should include facilities for matching the characteristics of the incoming signal and adjusting the picture. Specific details of how these parameters are changed should appear in the program documentation; however, Figs 1.3 to 1.6 show how incorrect settings affect the received picture.

Reception of satellite images

The satellites

There are two types of satellite that transmit weather images of interest to amateur observers. These are Geostationary Operational Environmental Satellites (GOES) and Polar Orbiting Environmental Satellites (POES). Geostationary satellites orbit at the same speed and direction as the Earth rotates, and so maintain a constant position above a single point on the Earth's surface. They are placed in high orbits, typically around 36 000 kilometres, which provides a view of almost a whole hemisphere of the Earth's surface. In contrast, at 800–1000 kilometres, polar orbiters have a smaller but closer view (see Fig 1.8).

As their name suggests, polar orbiting satellites travel in orbits that pass

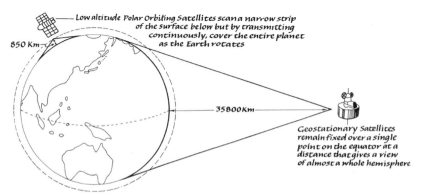

Fig 1.8 Relative coverage of GOES and POES satellites.

close to the Poles. The US NOAA satellites are a good example. Orbiting in the opposite (retrograde) direction to which the Earth rotates, they execute about 14.2 orbits per day, crossing the Equator at an angle of around 98 degrees. These are known as sun synchronous orbits, and pass over the same part of the Earth at approximately the same time each day.

Transmission characteristics

Again, taking the NOAA satellites as an example, GOES satellites make periodic scans of their coverage area which are transmitted back to Earth on a frequency of 1.7 GHz. The image is re-formatted adding coastal and political boundaries, and quickly transmitted back to the satellite for re-broadcast in a form that's more acceptable to professional users. However, reception requires a 3 metre dish antenna and is beyond the resources of most amateurs. GOES satellites also transmit the same image in a form known as WEFAX, but unfortunately this is not real time and may be delayed by up to three hours. This is because the satellites are also used to re-broadcast weather information and images from other sources.

WEFAX transmissions are also in the microwave region and require a special radio or frequency converter plus a directional antenna such as a 3 foot dish or Yagi. While some amateurs have been able to receive good-quality images without great difficulties, reception of VHF signals from polar orbiters is certainly less demanding and so is covered in more detail here.

Polar orbiting satellites also transmit in the 1.7 GHz band. These are high-resolution pictures (HRPT) but, in addition, they also transmit low-resolution pictures on frequency modulated VHF in a format known as APT (automatic picture transmission). Here, images are sent at a slower rate; 120 lines per minute, giving an effective data rate of 32K bits/second. Because signal strengths are quite strong, reception is relatively easy and the equipment less critical.

Geostationary satellite type	Frequencies		
NOAA	137.500	137.620	
Meteor	137.300	137.400	137.85

Table 1.2 Frequencies used by polar orbiting satellites.

As they travel between Poles, the satellite's cameras scan the Earth below in an east/westerly direction, producing a continuous picture band that slightly overlaps the previous orbit, and in this way they cover the entire globe. For Earth-bound observers, reception is only possible when the signal is above their horizon, but pictures obtained are of conditions as they actually are at the time of reception. Compare this to WEFAX GOES images that may be hours old, or some Internet sources that may only be updated once a day or less.

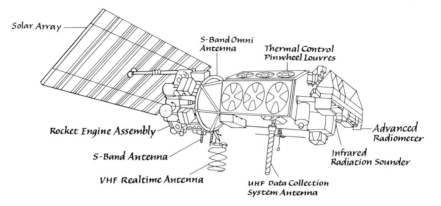

Fig 1.9 TIROS N polar orbiting satellite.

Equipment for receiving images from polar orbiters

The equipment needed includes a suitable antenna, radio, computer, and software for predicting when satellites will be in range; 137 MHz is a little above the FM local radio broadcast and air traffic bands, and a little below the amateur and marine bands. However, it is often covered by amateur multi-band transceivers and wide band frequency scanning receivers of the type used by 'enthusiasts' for monitoring police, ambulance, aircraft and cell phone conversations. Unfortunately, and here's the rub, the required bandwidth is 42 MHz, which is rather wider than that of most communications receivers and narrower than that provided by broadcast radios. With too much bandwidth, the signal becomes diluted with noise and clear reception becomes impossible. With too little, white areas on received images tend to appear grey – although for some purposes this may be acceptable.

From time to time articles have appeared on the Internet and in several radio/electronics magazines giving plans for the construction of suitable receivers and instructions for modifications to commercial scanners. The alternative, and the most trouble-free option, would be to buy a commercial radio that's purpose-built for APT reception, but whatever option you choose a search through the websites listed in Appendix 8 should provide the necessary information or put you in touch with suppliers.

The antenna

Because VHF propagation is essentially line of sight, reception of signals from polar orbiting satellites is only possible when the satellite is above your horizon. There are two requirements here. First, you'll need to know at what time of the day a particular satellite is within range, which is the purpose of the prediction software described in the next section. Secondly, you'll either need a tracking system and antenna that can be directed at the satellite as it passes overhead or you'll need an antenna that's able to receive signals from any horizontal or vertical direction. For this introduction to satellites, we'll look only at the second type, which is technically simpler and more suited to marine installations.

POES satellite signals are occasionally strong enough to be heard on a simple whip antenna as fitted to most hand held receivers, but the arrangement is far from ideal. A better arrangement is to use an antenna that's designed specifically for the purpose. Several types that have been used to good effect include the turnstile, lindenblad and quadrifilar. Of these, quadrifilar is probably the most widely used. To quote M Walter Maxwell, one of its developers: 'It comprises two bifilar helical loops oriented in mutual orthogonal relationship on a common axis. The terminals of each loop are fed in antiphase and the currents in the two loops are in phase quadrature.' Fortunately, you don't need to understand exactly how they work to use one effectively, or even to build one yourself. Plans are readily available from several Internet sites, and some manufacturers are listed in Appendix 8.

Satellite image software

Two essential pieces of software for receiving satellite images are a satellite tracker and an image display program. We'll look at these two items in turn.

During daylight hours, on any particular day there may only be a small number of operational satellites crossing your horizon, and a typical pass may last only 10 minutes. With a receiver that can scan through the five frequencies listed in Table 1.2, you

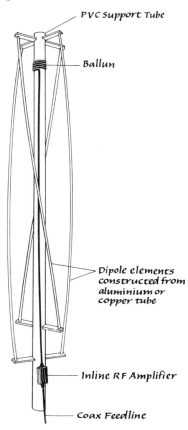

Fig 1.10 Quadrifilar antenna construction.

could simply listen all day until a signal from a passing satellite is heard. A better approach is to plan your listening by computing in advance the times at which particular satellites will come within range. This is the purpose of the tracker program, which can also compute a host of other useful data such as how long the pass will last, give an idea of signal quality, and perhaps show a ground 'footprint' to indicate the geographical extent of the picture you'll receive.

To track any satellite, some basic data is needed to describe its orbit, position and speed. These are provided in a set of figures called Kepler parameters after Johann Kepler (1571–1630), who made a thorough study of orbiting objects (see Appendix 7). Parameter sets for polar orbiters are obtainable direct from NORAD (see Appendix 8) and are presented in a plain text format that's acceptable to most tracking programs. Unfortunately, the orbits of satellites, particularly those closer to the Earth, are subject to small changes, and Kepler parameters change slowly with time. As a general rule, any that are more than 30 days old should be updated before use. Fig 1.11 shows a typical general-purpose tracking program that can be downloaded. The program (written by the author) has the usual ability to predict times of signal acquisition and loss, can plot several satellites simultaneously, and can show ground footprints and other views.

Software for displaying received images

Many of the software programs for receiving HF weatherfax can also be used to receive satellite images. Particularly popular are JVComm32 (shareware) and WXSAT (freeware). Sources of both are listed in Appendix 8. Since both use a simple connection between the radio's audio output and sound card microphone input, installation could hardly be simpler. However, there are some configuration adjustments that need to be made so that reading and following through the help file notes is the key to success.

Satellite reception software usually has the capability of saving images in a graphics format such as a bitmap, JPG or GIF file. The software may also have some ability to make modifications – perhaps to trim the edges or remove interference. These functions are usually quite limited, so another desirable piece of software to have would be a general-purpose image file editor of the type used to edit photos or to prepare webpage illustrations. Some useful adjustments you might make to received images could include changes to the size or contrast, though best results require a good deal of trial and error and, until you're familiar with the process, it's possible to spend huge amounts of time without achieving significant improvements.

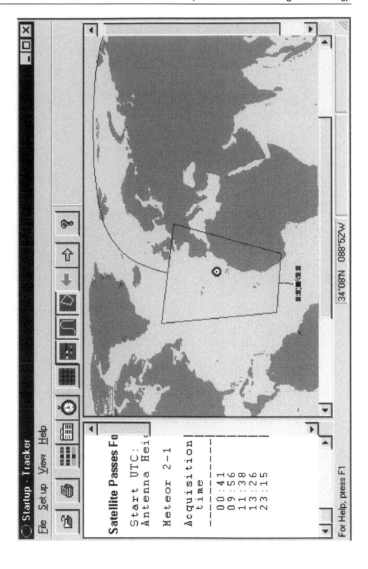

Fig 1.11 Tracker – a general-purpose satellite tracking program that can be downloaded from www.pangolin.co.nz.

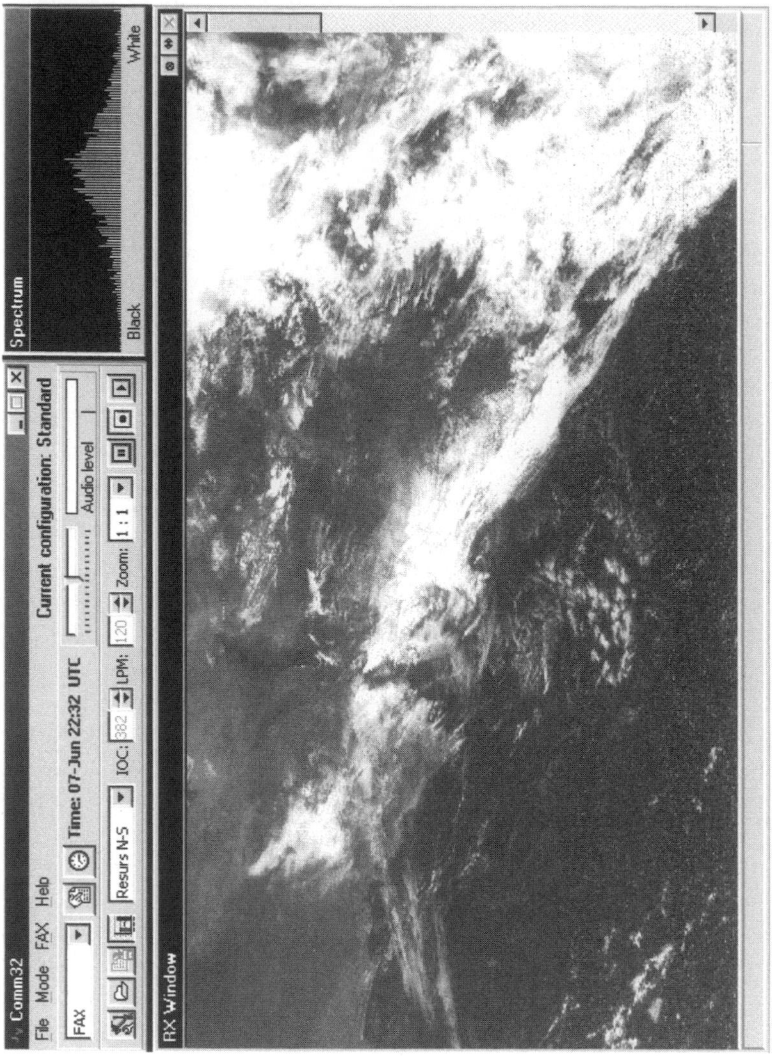

Fig 1.12 JVComm32 displays an image from a polar orbiter.

Other sources of satellite images

Some HF weatherfax stations, notably those run by the US Coast Guard, but also Honolulu and Delhi Meteo, include re-broadcast images from geostationary satellites. These are transmitted with an 8 bit grey scale and are an easy option if you have no special satellite antenna and receiver, but are already set up for receiving HF fax (see Fig 3.9). However, at the time of reception they could be several hours old as there may be some time lapse between reception of the image and the time of re-broadcast at the appropriate slot in the station's schedule.

The Internet is another major source, where most of the world's weather bureaux publish a huge range of images with many special views – some colour enhanced and designed to show particular aspects of atmospheric conditions. Make sure that you check their validity dates as some sites are not updated regularly, and show views that may be months old. Direct images from polar orbiting satellites have the great advantage that they're transmitted in real time. The image that you see develop on your computer screen is of weather as it actually is at that moment.

For the future

The APT technology used to transmit low-resolution images is several decades old, and it is by no means sure how long the current service will remain available. Encryption of transmissions is a possibility that has been suggested, and a gradual transition from APT to the low-resolution picture transmission (LRPT) format has been announced by the WMO.

If the long-term future for amateur reception of images from weather satellites is unclear, it's easier to be more confident about the Internet. For boats at sea, and those based in remote parts, an Internet connection of sufficient speed to receive pictures is still difficult and expensive to arrange, but it seems certain that in the not too distant future this will change. Whether faster and cheaper wireless connections will be provided by satellite or terrestrial cell phone-type services matters little as, in either case, received images would be displayed on an already familiar web browser. Gone would be the need to learn about special software and tracker programs, but instead expect service charges, and to be a stage removed from the actual data source.

Chapter 2

World Weather Patterns

I f you are hove-to in a gale in the English Channel and feeling sick, what do you care about trade winds in the Caribbean or the kind of winds that airline passengers are experiencing at 10 000 metres? Of course, it is only the weather that you are actually going to get that really counts, but local conditions are the inevitable product of those occurring elsewhere. The weather is everywhere. There are no glasshouses where conditions are isolated from what is happening next door.

Often, quite small local events can trigger the build-up or decay of large weather systems, affecting, maybe, a whole continent. As a background to understanding local weather patterns, it is important to know a little of how air masses are circulated globally, and this is the main topic for this chapter. In detail, the problems of atmospheric modelling are extraordinarily complex, and a challenging application even for the world's largest computers, but our purpose here is a little less ambitious. We begin by looking at the main forces behind world weather and see how they help to account for the weather patterns found in the tropics, temperate and polar regions.

Forces driving the weather

Heat from the sun, or rather the temperature difference it creates, is the main energy source behind world weather. In polar regions, particularly in winter when the earth's axis is inclined away from the sun, precious little solar radiation strikes the Earth's surface and extremely low temperatures are the result. Compare this with the tropics, where the sun passes immediately overhead. Radiation density over land and sea is maximised and higher surface temperatures are the result.

Air masses of different temperatures have different densities and, rather like oil and water, do not mix easily. Converging warm and cold air masses clash, and swirl around each other causing turbulent conditions that may lead to disturbed weather such as squalls, gales or storms.

As surface air near the Equator is warmed, it expands. It becomes lighter, more buoyant and rises, creating a lower pressure region, called the equatorial trough, that pulls in cooler air which is in turn heated and made to rise. In the absence of other influences, winds would blow along

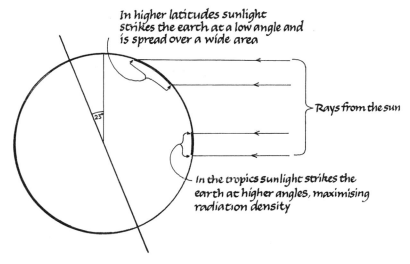

Fig 2.1 Contrasting densities of solar radiation in tropical and polar regions.

straight paths, travelling directly from areas of high pressure to lower pressures. However, this simple idea is modified by the Coriolis[1] force, which originates from the Earth's rotation through space. Its effect is to push winds a little to the right in the northern hemisphere and to the left in the south. As a result, winds travelling from mid latitudes towards the equatorial trough are deflected in a westerly direction, so forming the north and southeasterly trades. Actually at the Equator the Coriolis force is nil, but it increases with latitude and causes winds to spiral around pressure systems in other parts of the world.

The tropics

For our purposes, it is convenient to think of the tropics as extending between latitudes 30° north and 30° south, which is an enormous area covering half the surface of the globe. Dominated by trade winds, it has some of the most stable and predictable weather. Long periods of calm or gales are unusual. Trade wind belts vary in strength between seasons, the strongest being in winter when they extend further towards the Poles. Also found in the tropics are monsoon areas of the northern Indian Ocean, north-west Pacific and West African coasts. Here the trades are modified by the seasonal heating of continental land masses, and in the summer months their direction is altered or even reversed.

---

[1]Coriolis force is proportional to wind velocity and the sine of the latitude. It is named after the French physicist Gaspard Coriolis (1792–1843).

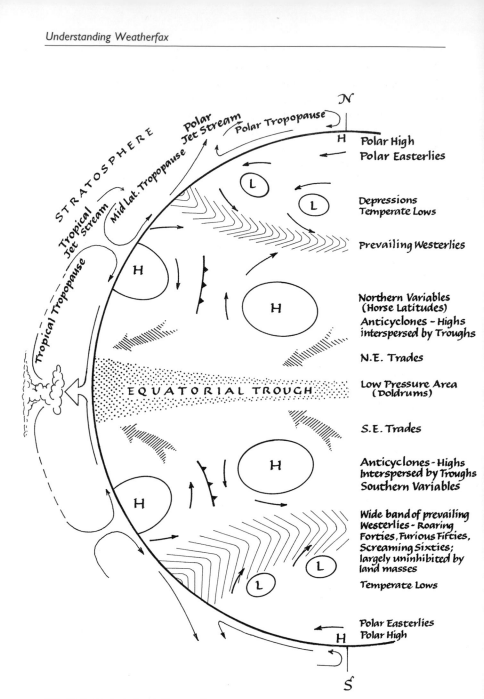

Fig 2.2 Global circulation patterns.

The inter tropical convergence zone (ITCZ)

Trade winds of the two hemispheres converge near the Equator at the ITCZ, or equatorial trough, which is a band of low surface pressure and ascending airs. It varies in width and is characterised by humid conditions and light variable winds. Sailing vessels passing through the ITCZ, or Doldrums, as it was known, often experience difficult conditions. Localised squalls are a common occurrence and often bring strong winds from any direction with the possibility of heavy rain showers. These can be a dangerous trap as one moment you may have all sail set trying to catch the lightest breath of wind and then, when it hits, you are blasted by strong winds with the risk of damage to sails and rigging. Squalls are usually too small to appear on weather charts and the barometer gives no indication of their approach, but it is as well to keep a good look-out and be prepared to reduce sail quickly. Warning signs are a sharply defined, heavy build-up of cumulonimbus, though this is not always easy to spot at night. Typically, squalls are over within half an hour.

The ITCZ is also a starting point for the development of rotational tropical storms (ie tropical cyclones or hurricanes; see Chapter 6). In the north-west Pacific, from July to October, these develop at the rate of about one per week, and tend initially to move westwards along the ITCZ before recurving. Summer months are the storm season for all oceans, but – storms apart – this is normally a time of light winds and weak trades.

Polar regions

Polar latitudes above 60° account for only one-seventh of the world's surface area, but are the main source of cold airs driving the weather in other parts of the planet. Cold descending airs form regions of high surface pressure at the Poles and surface winds known as polar easterlies.

Rossby or hemispheric waves

At a height of 6–15 km, high-speed winds circulating around the planet tend to flow not along straight paths but in enormous curves, forming lobes or waves. These are known as Rossby waves and their formation often follows a four- to six-week cycle of growth and decay (see Fig 2.3). At any moment, some are decaying while others are building, though all are generally moving in an easterly direction. As an individual wave grows, it may spawn counter-rotating eddies that eventually break away from the initial air stream and drift off towards lower latitudes.

At the surface, these waves tend to overlay an alternating sequence of high and low pressure systems that are a prominent feature of temperate-zone weather. For forecasters, upper level wind directions are significant as they are a steering force for surface weather and provide an important indication of the direction in which particular systems will move.

31

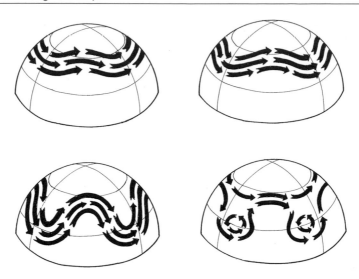

Fig 2.3 Development of east-going upper air waves around the Poles (in this case the North Pole).

Mid latitudes

Mid latitudes, between the tropics and polar regions, cover about a third of the Earth's surface and are the region where most cold/warm-air conflicts occur. The underlying wind direction is from the west and in the southern hemisphere these winds are particularly strong, hence the popular names: 'roaring forties', 'furious fifties' or 'screaming sixties'. However, in either hemisphere these winds have nothing like the consistency of the trades. The constant succession of east-going low pressure systems causes winds to routinely shift through all directions and bring regular gales and rain. The sequence in Fig 2.4 shows how warm and cold fronts are formed at the interface of air streams of differing temperature. Fronts are of special interest to forecasters as they are often associated with wind shifts (changes in direction), increasing cloud and rain.

Fig 2.5a shows the same developing warm and cold fronts, but this time with isobars added. These are lines linking points of equal pressure rather like the way contours on a land map link points of equal height. Isobars are widely used on many types of weather maps and indicate the spread of pressure across the chart. The unit used today is the hecto Pascal, which is equal to 100 Pascals and equivalent to the older Millibar (Mb).

Figs 2.5a, b and c also show further development of these fronts as they begin to form an active surface low pressure system or depression. They are a dominant feature of mid-latitude weather and when below about 1000 hPa can lead to increased wind, gales or even storms if the low increases in depth.

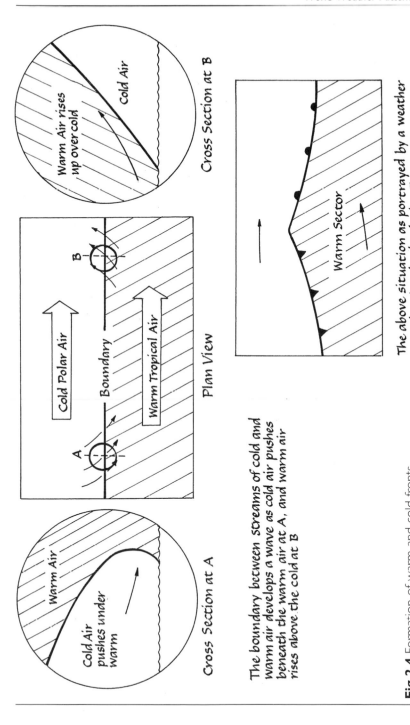

Cross Section at B

Warm Air rises up over cold

Cold Air

Plan View

Cold Polar Air

Boundary

Warm Tropical Air

B

A

Cross Section at A

Warm Air

Cold Air pushes under warm

The above situation as portrayed by a weather map, showing the developing Fronts

Warm Sector

The boundary between streams of cold and warm air develops a wave as cold air pushes beneath the warm air at A, and warm air rises above the cold at B

Fig 2.4 Formation of warm and cold fronts.

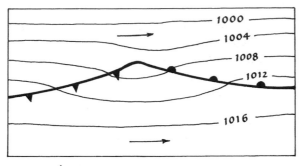

a. Developing Wave   see FIG.2.4

Fig 2.5
Development of a pressure system (northern hemisphere).
 Diameters range from 100 to 2000 miles, and typical life spans are from 4 to 6 days – after which they lose their identity by regaining pressure and filling.

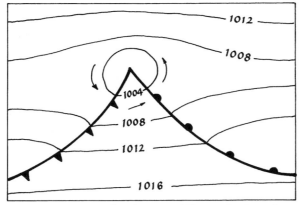

b. Low Pressure develops rapidly at the junction of Warm and Cold Fronts

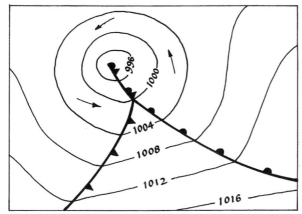

c. The Depression continues to deepen with an Occluded Front forming as the Cold Front catches up with the Warm

Cloud formation

Air is able to hold water vapour in a similar way to that in which water is able to dissolve solids like sugar, though it is perhaps unusual to think of wind drying wet clothes by dissolving the water they contain. If you keep adding sugar to a saucepan of water there comes a point after which the water becomes saturated and will not accept more. Unless, that is, you warm the water a little, raising the saturation point and allowing more to dissolve. Similarly, water can be dissolved in air to form a completely transparent gas. At any temperature there is a maximum concentration or saturation point above which, like a sponge, a fixed volume of air can hold no more. If such saturated air is cooled a little, the excess water precipitates out from vapour to liquid and appears as finely suspended droplets forming clouds or fog. The temperature at which this begins to happen is known as the *dew point*, and if the air is further cooled, droplets amalgamate and fall as rain.

When warm moist air flows are forced to rise, as they are when they encounter mountain ranges or a heavy cold air mass, as in a frontal system, cooling takes place and large amounts of water are precipitated to form clouds. Their shape and form can differ enormously and give important clues to the severity and intensity of the weather system that produced them. These are topics we'll examine more closely in Chapter 4.

Cloud views from space

Clouds are also the most important meteorological features in satellite images. Because the Earth's atmosphere is largely transparent, its presence is only shown when it holds particles such as water or ice that can absorb, reflect or refract sunlight. Although only the tops of clouds are seen, they still provide a useful guide to current and future weather.

Recognisable patterns include the development of the comma-shaped cloud around low pressure systems and which, depending on conditions, may dissipate or further develop into the broken doughnut shape of a severe storm system.

When the sun is at an oblique angle, dense clouds can throw shadows, either on lower clouds or the Earth's surface, and by measuring their length it is possible to estimate cloud heights. Tall clouds indicate regions of strong convection, and are a common feature of the ITCZ. Look out for small peaks projecting above an expanse of thick, flat-topped cloud. These are known as *overshoots* and are indicators of convection cells and possible thunderstorm activity.

Chapter 3

Charts and Images: Types and Features

Weatherfax stations broadcast a huge variety of different types of charts and pictures. Some are general-purpose charts; some are tailored to meet the needs of particular groups of users such as fishermen, aviators or farmers. Others may be concerned with any special characteristics of the coverage area, such as the presence of icebergs, fog or currents. Every station produces a schedule or list of its fax products with their time of transmission. These are collectively published in frequency list references (see page 100), though many stations include the schedule as one of their daily broadcasts. Table 3.1 shows a schedule for the United Kingdom station GYA.

On the whole, most users find that fax stations produce far more information than they can comfortably handle, so it is necessary to look carefully at the schedule and plan reception to receive only the transmissions that are most useful. Unfortunately, schedules are often changed, and even official lists can be out of date by the time they are sold. Some users get over the problem by running the receiver for a whole day and collecting every fax that is put out. With a full 24-hour output you can gather the best idea of exactly what fax products are available and, by comparing the quality of those that are transmitted more than once, you can plan your future listening for times when reception is at its best. With possibly as many as 48 charts to print, a conventional weatherfax machine would use a considerable amount of paper. Using a computer with software able to store images on disk is a more convenient option, particularly if several pictures can be compared within separate windows on the same screen.

In the remainder of this chapter, we look at examples of the most generally useful type of fax charts and the kind of information they provide. Though these are taken from around the world, except for the fact that air circulates around lows and highs in opposite directions in the two hemispheres, there is little utilitarian difference between, say, a surface

TIME	CONTENTS OF TRANSMISSION	RPM/IOC	VALID TIME
1000/-----	Surface prog T+24	120/576	0600
1012/-----	Spot winds 500MB T+24	120/576	0000
1024/-----	Spot winds 400MB T+24	120/576	0000
1036/-----	Spot winds 300MB T+24	120/576	0000
1048/-----	Spot winds 250MB T+24	120/576	0000
1100/-----	Surface analysis	120/576	0600
1124/-----	Fronts centres winds T+96	120/576	0000
1136/-----	Fronts centres winds T+120	120/576	0000
1148/-----	Gale summary	120/576	1200
1200/-----	Surface analysis	120/576	0600
1212/-----	SST 10 per cent ice edge	120/576	Mon/Tue
1224/-----	Layer depth	120/576	Tuc
1236/-----	CZ potential	120/576	Tue
1248/-----	Minimum sound channel depth	120/576	Tue
1300/-----	Surface prog T+24	120/576	0600
1312/-----	Ship ice accretion 0 deg C level T+24	120/576	0000
1324/-----	Poor visibility T+24	120/576	0000
1400/-----	Fronts centres wind T+72	120/576	0000
1436/-----	Schedule	120/576	
1500/-----	Surface analysis	120/576	1200
1512/-----	Scexa tafs	120/576	1500
1524/-----	Scexa tafs continued	120/576	1500
1536/-----	Frontal positions	120/576	Thu
1548/-----	Gale summary	120/576	1600
1600/-----	Fronts centres winds T+48	120/576	0000
1700/-----	Fronts centres winds T+48	120/576	0000
1724/-----	500MB height T+0	120/576	1200
1736/-----	Surface prog T+24	120/576	1200
1748/-----	500MB height T+0	120/576	1200
1800/-----	Surface analysis	120/576	1200
1812/-----	300MB height T+0	120/576	1200
1824/-----	300MB height T+24	120/576	1200
1836/-----	500/1000 thickness T+0	120/576	1200
1848/-----	500/1000 thickness T+24	120/576	1200
1900/-----	Gale summary	120/576	1900
1912/-----	Sea swell T+24	120/576	1200
1924/-----	850MB WBPT T+24	120/576	1200
2000/-----	Fronts centres wind T+48	120/576	1200
2012/-----	Fronts centres wind T+72	120/576	1200
2024/-----	Fronts centres wind T+69	120/576	1200
2036/-----	Fronts centres wind T+120	120/576	1200
2100/-----	Surface analysis	120/576	1800
2112/-----	Sig wind areas T+24	120/576	1200
2124/-----	Sig wind areas T+48	120/576	1200
2136/-----	Sig wind areas T+72	120/576	1200
2148/-----	Sig wind areas T+96	120/576	1200
2200/-----	Surface prog T+24	120/576	1200

Table 3.1 Transmission schedule extract for Northwood (UK station GYA April 2001).

analysis from Australia, the USA or the UK. Plotting symbols for isobars and fronts have been internationally standardised, and weather changes associated with the passage of lows, highs and fronts have much in common.

The surface analysis

Sometimes called the mean sea level analysis (MSL ANAL), this is the familiar synoptic chart showing isobars, fronts and pressure systems that are shown in newspaper and television forecasts. Fig 3.1 is a typical example covering the South Pacific and is transmitted by station ZKLF from Auckland, New Zealand.

The chart is compiled from data collected from a variety of sources, including land and ship observation stations, drifting buoys with automatic reporting equipment, satellite images and weather radar pictures. Its most obvious feature are the isobar lines that appear closely spaced near the bottom of the chart, form whorls in the mid latitudes and are hardly present within the tropics. On this particular chart, they are drawn at 5 hPa intervals and indicate the pressure to be expected at sea level. Incidentally, where these pass over mountain ranges, pressures are corrected to sea level values as if the land was removed. Major high and low pressure systems are marked with a letter **H** or **L**, though even without the letter it would still be easy to see which were which from the surrounding isobars that show increasing or decreasing pressure towards the centre. Some are marked with a centre pressure and have an arrow showing their direction of movement and speed in knots. The chart also shows a typical succession of alternating highs and lows moving east between latitudes 20° to 70°, similar to those that sweep similar latitudes in the northern hemisphere.

Associated with low pressure systems are a series of fronts shown as heavy black lines marked with semicircles or triangles indicating regions where cold and warm airstreams meet and often bring disturbed weather. Less obvious but also of significance are high and low pressure ridges and troughs. Again, these can be picked out by imagining the isobars to be like land contours with highs and lows appearing as hills and dents.

Finally, an important point to note when considering any chart is to check the map projection to which it has been drawn (see Appendix 3). Most navigational charts use Mercators, but Fig 3.1 is polar stereographic. For the area covered, this has the advantage that land areas are less distorted, but notice that east/west courses are plotted as curves centred on the Pole that is off the chart. North/south courses are straight lines, but radiate from the Pole.

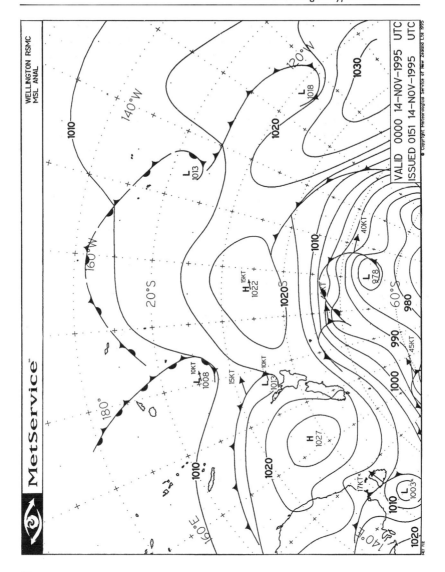

Fig 3.1 Mean sea level analysis for the South Pacific.

The mean sea level prognosis (MSL PROG)

Fig 3.2 shows an example of a mean sea level prognosis. This chart was actually transmitted at 2345 UTC on 13 November and was prepared (issued) at 2211 UTC on the same day, a few hours earlier than the chart shown in Fig 3.1. The important difference here is that, since it is a prognosis, the isobars and other features are placed in the positions they are expected to occupy some 30 hours later. Its predictions are based upon computer models of global and local atmospheric conditions, tempered with a final manual interpretation based on the experience of the forecaster.

Chaos theory and the 'butterfly' effect

Even with the aid of the most powerful computers, atmospheric modelling is debatably more of an art than an exact science. Often, a seemingly insignificant event can initiate a diverging series of events with consequences far removed in time and space from the initial trigger – rather like the discarded cigarette that ignites a few dead leaves, that burns a forest, that makes smoke that blows over a town, that blackens the washing, that sells more soap powder, that . . .

The weather contains many such 'cigarette ends' or conditions with a potential for promoting or inhibiting the development of much larger weather systems. Given sufficient detail of the current state of the atmosphere, it is possible to predict with reasonable accuracy the way in which the weather will evolve over the coming hours. However, as the time scale is increased to several days, the scenario becomes increasingly susceptible to random unforeseeable 'cigarette end'-type changes. It becomes impossible to analyse by traditional mathematical methods. Instead, a new branch of mathematics (chaos theory) has emerged and is concerned with the study of such systems, that in addition to the weather includes applications in fields as diverse as economics forecasting and cardiology.

Use of the prog chart

A comparison of the surface analysis with the surface prognosis gives the best indication of the way in which the weather pattern is expected to evolve. When planning weather tactics it is useful to know the directions in which frontal systems, low and high pressure areas etc, are likely to move and if they are expected to weaken or become more vigorous. For this reason, the prog and anal charts are the barest minimum needed to make a useful assessment of weather developments.

Upper air charts

We live mainly on the surface of the planet, so, unless you are the pilot of an aircraft, what use is there in knowing what the weather is doing a

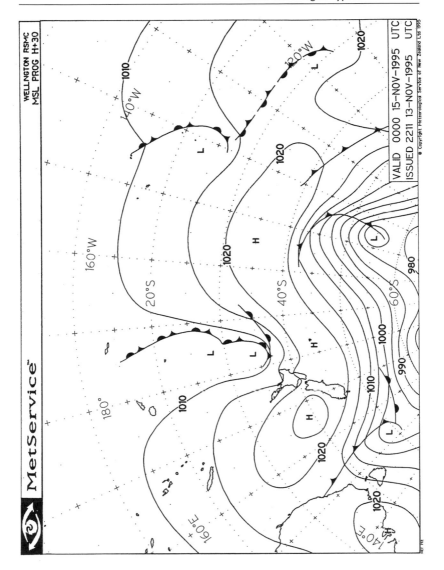

Fig. 3.2 Mean sea level prognosis for the South Pacific.

kilometre or more above our heads? In reality, weather is a 3-dimensional phenomenon. Rain comes from clouds that are blown about by the upper winds or formed by high altitude temperature changes. Several cloud-producing features appear clearly on upper air charts, but may not appear at all on the surface analysis.

Barometric pressure reduces with altitude at a rate of about 1 hPa per 100 metres. An easy way to put this to the test is to take your barometer along to a high building with a lift and note the difference in reading between the ground floor and the roof. Fig 3.3 shows an upper air chart prepared for a height where the barometric pressure is 500 hPa.

At first sight, the analysis chart in Fig 3.3 appears much like a surface analysis; however, do not be deceived. Instead of isobars, the lines circling the low and sweeping the bottom of the chart are, in fact, height contours labelled in metres above sea level. Like a land map, the chart shows a surface, but an imaginary one at heights between 4.9 and 5.9 kilometres where the barometric pressure is 500 hPa. The 'low' in the middle is like a dent in the surface and, as with any northern hemisphere low, the wind arrows show winds circulating in an anti-clockwise direction. As well as direction, wind arrows also indicate strength. Each full-length tail barb represents a speed of 10 knots. Half-length tail barbs are 5 knots and black triangles 50 knots (see Appendix 4). Heavy dashed lines on the chart indicate low pressure troughs.

Some upper air charts, notably those from New Zealand, have no height contours, but instead are overprinted with a grid of wind arrows that also include air temperatures in degrees Centigrade. This can be a little confusing as temperatures are mainly negative, but are printed without a minus sign. Once again, the positions of highs and lows and other features are easily determined by observing the direction of wind flows; also, low pressure systems usually have temperatures lower than their surroundings. Upper air analysis and prognosis charts are prepared for a range of pressures between 850 and 200 hPa (as shown in Table 3.2). How they are used to help forecast surface weather is a subject for the next chapter.

Other charts

In addition to the basic analysis and prognosis charts described so far, most fax stations produce a number of charts for users with special interests, eg forecasters, fishermen, foresters, flyers or farmers. It would be impractical to provide a complete list of every type available, particularly as broadcasters are continually upgrading and extending their products to meet changing needs. The illustrations that follow (Figs 3.4–3.7) are a few examples of special-purpose charts for mariners.

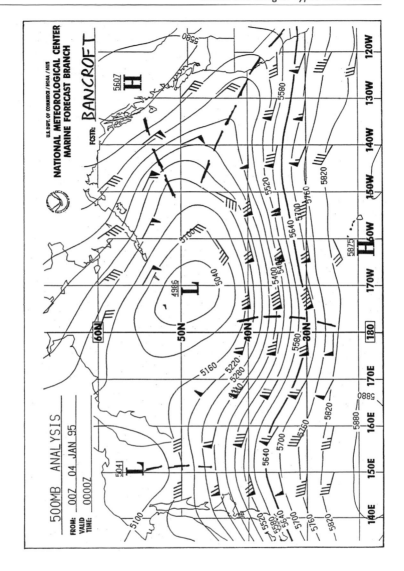

Fig 3.3 500 hPa analysis chart covering the North Pacific.

Pressure level (hPa)	Base height	Flight level (feet x 100)	Contour spacing (metres)
850	1470	FL 50	30
700	3000	FL 100	60
500	5580	FL 180	60
400	7200	FL 240	120
300	9120	FL 300	120
200	11820	FL 390	120

Table 3.2   Upper level chart heights and pressures. Flight levels are in common use on charts from the USA and Europe.

Significant weather features (SIG WX)

This is another chart intended mainly for aviators (Fig 3.4), but it is also of interest to mariners because the clouds shown are those associated with active thunder showers, ie cumulonimbus (see Chapter 4). In the Doldrums, these are areas of squalls – though isolated squalls are not identified. Abbreviations used are as follows:

ISOL EMBD CB   Isolated individual cumulonimbus clouds which are embedded in an area of other layered cloud.
OCNL EMBD CB   Occasional well-separated cumulonimbus clouds which are embedded in an area of other layered cloud.
FREQ CB   Frequent cumulonimbus clouds with little or no separation.
FREQ EMBD CB   Frequent cumulonimbus clouds with little or no separation which are embedded in an area of other layered cloud.

Heights of significant weather are given in flying levels (FL) suffixed by a height in hundreds of feet. For example:

320   Refers to a feature whose top is at 32 000 feet and whose
260   base is at 26 000 feet. If the feature extends above or below the layer of atmosphere covered by the chart, its height is recorded as XXX.

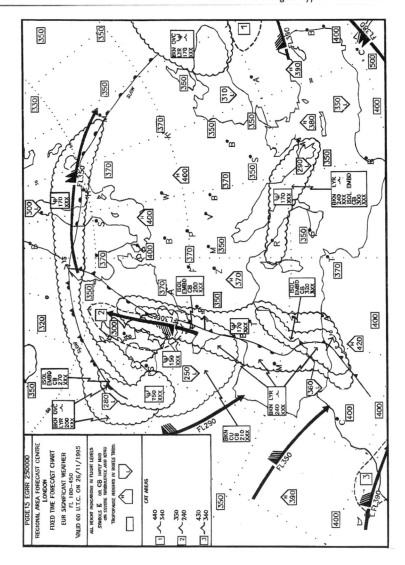

Fig 3.4 Significant weather features (SIG WX).

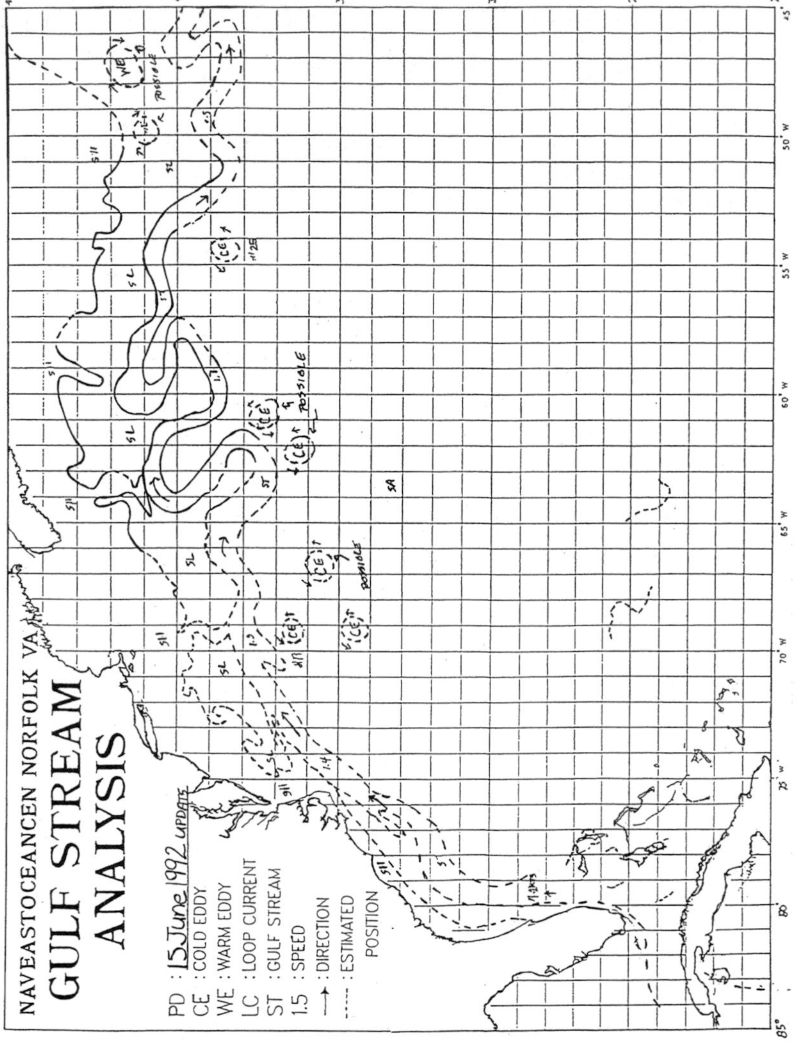

Fig 3.5 Gulf Stream analysis. Originating in mid-latitudes of the W Atlantic, warm Gulf Stream currents flow NE across the Atlantic to warm the W coasts of Ireland, the UK and Norway. The stream can reach speeds of several knots and frequently meanders into large loops and eddies that may remain in place for days. It's sometimes described as a 'river in the ocean' with currents of 2 knots or more, sharply defined boundaries and rough seas where it opposes the wind.

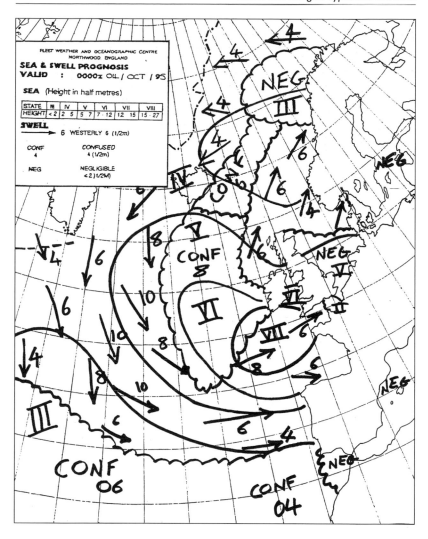

Fig 3.6 Sea swell prognosis. If you are in a sheltered harbour contemplating an offshore passage, it is often useful to know what kind of sea conditions to expect. The swell prognosis can provide a useful guide though is of little use close inshore where refraction and reflection effects can cause large changes in wave size and direction. Where large waves are indicated, they often persist for several days affecting regions well beyond their area of formation.

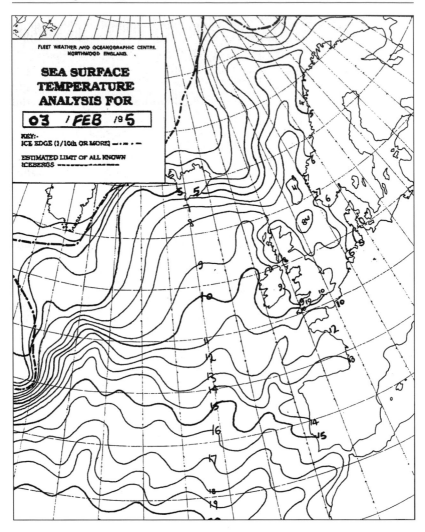

Fig 3.7 Sea surface temperature analysis.

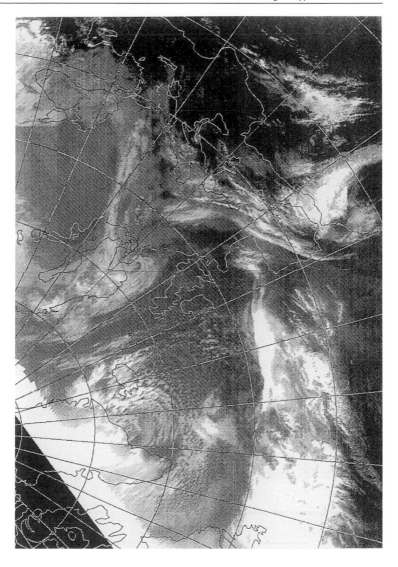

Fig 3.8 Re-broadcast GOES satellite image of the north Atlantic and western Europe.

Satellite imagery

Satellite cameras are 'tuned' to respond to light from a particular band of wavelengths. This may be visible light (VIS), where the image produced is the same as the view that you'd see if you were actually aboard the spacecraft. VIS images can only be obtained from parts of the Earth that are in sunlight, and at night some satellites switch to a camera that is sensitive to a part of the infra red (IR) spectrum. These images are essentially a 'heat picture' of the Earth's surface and cloud tops, showing the coldest regions as white, and hottest parts as black. Intermediate temperatures appear as shades of grey between these two extremes. IR images are also useful during the day and convey a different set of information from the VIS. Table 3.3 lists some of their different characteristics.

Visible light (VIS)	Infra red (IR)
Seas and lakes appear black, except when smooth, and reflect sunlight towards the satellite (ie sunglint, see pages 66–7).	Contrast between land and sea areas is usually good, but depends upon their relative temperatures. Contrasts are lowest in early morning when rapid night cooling may cause land to appear lighter, but improve as the land heats up to maximum temperatures in mid-afternoon. High contrasts may indicate land or sea breezes (see page 76).
Parts of the Earth where the sun is overhead are brightest.	
Thick clouds appear as opaque white, and thinner clouds, medium grey.	Features of similar temperature may be difficult to distinguish – for example, low cloud over land, particularly when the land is frosted or snow covered.
Oblique sunlight causes tall cloud systems to cast shadows.	

Table 3.3   Characteristics of VIS and IR images.

Image enhancement

IR images from most satellites are transmitted as a grey scale using 256 shades (which is about 100 times more than the human eye can resolve) to cover the range of temperatures between 57°C and 109°C. Black is used for the highest temperature, and white the lowest. In unenhanced images a linear relationship is used to cover intermediate temperatures, as shown in Fig 3.9a (page 52). However, because temperatures of meteorological interest normally fall within the range minus –80°C to 40°C, sharper contrasts can be achieved by using the entire grey scale to cover just this section, as shown in Fig 3.9b. Other enhancements plans shown in b, c and d may also be encountered, and are used to highlight specific features.

By definition, IR images are of radiations that lie outside of the visible spectrum, and so colour can have no true meaning. None the less, it is often used to enhance images, particularly in the press, Internet and television. Here, colours are sometimes chosen to match the natural colours of highlighted features, though often they are used to pick out temperature-sensitive events and isolate temperature events, and visual effects can be quite unreal and garish.

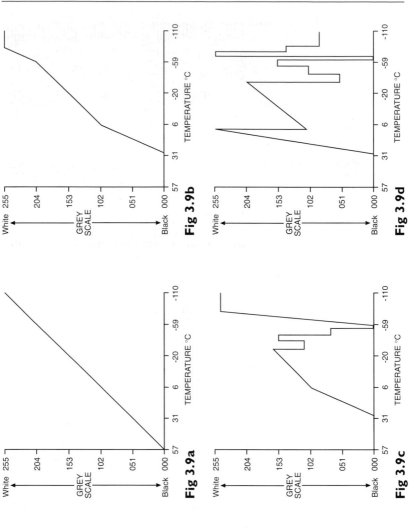

Fig 3.9 IR image grey scale temperature relationship curves. **a** Linear relationship. **b** General-purpose (ZA) enhancement. **c** Cloud top enhancement (TB). **d** Hurricane enhancement (BD).

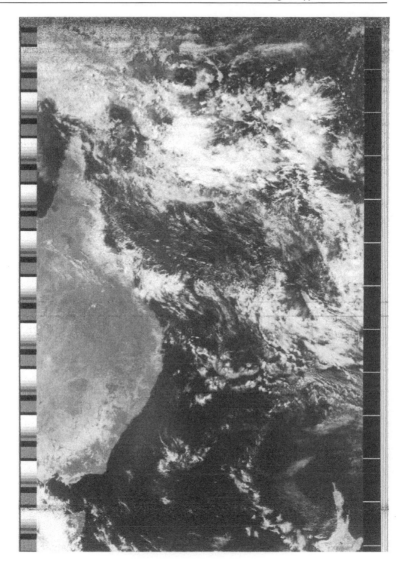

Fig 3.10 A north to south pass of NOAA 16. This VIS image of eastern Australia was received by a station in Brisbane. Picture quality is best when the satellite is overhead.

Chapter 4

Interpreting Details

With surface analysis and prognosis charts, you have a good picture of the present weather situation and the way that pressure systems, fronts, isobars etc are likely to change in the immediate future. But what do these features mean in terms of actual wind, rain, fog, snow or sunshine? In short, how do you read a weather map? In this chapter we examine some general principles and look more closely at chart details and the types of weather they are likely to produce.

Estimating wind speed

Winds are caused by differences in air pressure. The greater the difference, or, in other words, the steeper the pressure gradient, the stronger the wind. On weather charts the pressure gradient can be found by measuring the spacing between isobars, but in relating this to wind speed other factors need to be taken into account. These are:

- **Latitude** A given wind speed is denoted by a greater isobar spacing in the tropics than at the Poles.
- **Direction of isobar curvature** For a given isobar spacing and latitude, winds blowing round high pressure systems are stronger than those blowing round a low (cyclostrophic effect).

Geostrophic wind scales

Some fax charts include a scale for converting isobar spacing into wind speed. The print is often small and can be difficult to read if reception is poor. Fig 4.1 shows an example.

Geostrophic wind scales are only intended to be used on the chart on which they are printed. If the chart you are using does not have one, the following method, derived from a procedure from New Zealand MetService, is a useful alternative.

Step 1   Estimate the isobar spacing in nautical miles and the mean latitude of the point where you wish to measure wind speed. (Use the latitude scale 1° of latitude = 60 nautical miles.)

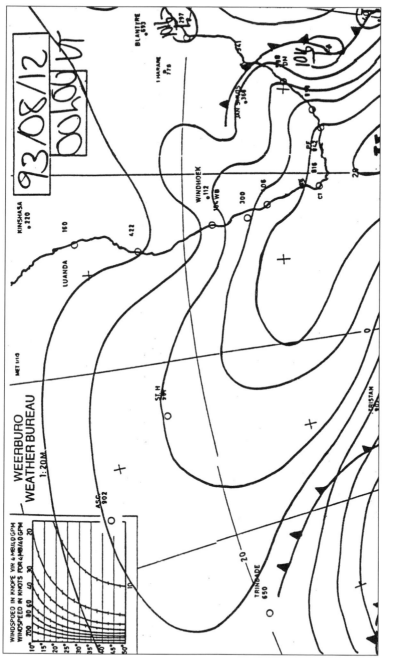

Fig 4.1 Geostrophic wind scale.

Estimating the direction of the wind

Buys Ballot's law

Instead of following the pressure gradient and blowing directly into or out of low or high pressure systems, winds are also subjected to Coriolis forces that cause them to follow a spiral path at a small angle to the direction of the isobars. From Fig 2.3 we can see that in the northern hemisphere their resultant direction around low pressure systems is anticlockwise and clockwise around highs. In the southern hemisphere these directions are reversed, but the relationship has several useful practical applications, so it is important to remember it correctly.

Stand facing the wind and, if it is due to a low pressure system, the centre will lie on your:

- left side if you are in the southern hemisphere (**L**ows, **L**eft, **L**ower hemisphere), and
- right side if you are in the northern hemisphere.

Alternatively, you can use your hands:

- Imagine a globe the familiar way up with the northern hemisphere uppermost.
- Place your left hand palm up or down to cover the hemisphere in which you are located.

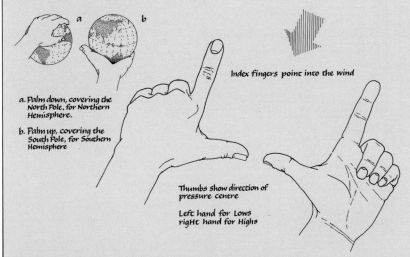

a. Palm down, covering the North Pole, for Northern Hemisphere.

b. Palm up, covering the South Pole, for Southern Hemisphere

Index fingers point into the wind

Thumbs show direction of pressure centre

Left hand for Lows right hand for Highs

Fig 4.2 Using the wind to find the direction of a low pressure system. Use your Left hand for Lows; the right works for highs.

- Use your index finger to point at the direction of the wind and your outstretched thumb indicates the direction of the low pressure system.
- Remember, your **L**eft hand for **L**ow pressure systems and **rigHt** hand for **H**ighs.

The above, of course, assumes that you know which hemisphere you are in, which is not usually a problem, and that you know the type of weather system that produced it. This is not usually a problem either, as both types of weather have different characteristics (see pages 58–61). The average surface barometric pressure taken over the Earth as a whole is 1013 hPa, so a further clue can be obtained by watching your barometer:

- Pressures of less than 1005 hPa are usually associated with lows.
- Pressures of more than 1020 hPa are usually associated with highs.

By applying Buys Ballot rotation to charted centres of high and low pressure you can also determine the direction of associated winds and how it is likely to shift as the weather system passes. Fig 4.3 shows wind directions associated with high and low pressure systems. Surface winds blow towards the centre of lows and outwards from highs, making an angle of about 20° to the direction of the isobar. This angle is due to surface frictional effects, and is a little more over land, but diminishes to zero at 6000 metres.

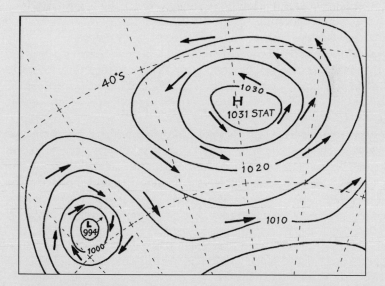

Fig 4.3 Surface winds associated with centres of high and low pressure. (The example is for the southern hemisphere.)

Step 2 Enter the distance in the first column of Table 4.1 under the isobar spacing used on your particular chart,[1] then read across to find the wind speed at your particular latitude.

Step 3 Correct for isobar curvature.
Increase wind speeds around a high by 20 per cent
Decrease wind speeds around a low by 20 per cent to 40 per cent (use higher reductions for higher speed winds blowing close to the centre)

Isobar spacing (nautical miles) 4 hPa 5 hPa	Speed (kt) 20° mean lat	Speed (kt) 30° mean lat	Speed (kt) 40° mean lat	Speed (kt) 50° mean lat
48 60	98 gusts to 146	67 gusts to 100	52 gusts to 78	44 gusts to 65
96 120	49 gusts to 73	33 gusts to 50	26 gusts to 39	22 gusts to 33
144 180	33 gusts to 49	22 gusts to 33	17 gusts to 26	15 gusts to 22
192 240	24 gusts to 37	17 gusts to 35	13 gusts to 19	11 gusts to 16
240 300	20 gusts to 29	13 gusts to 25	10 gusts to 16	9 gusts to 13
288 360	16 gusts to 24	11 gusts to 17	9 gusts to 13	7 gusts to 11

Table 4.1   Table relating wind speed in knots to latitude and isobar spacing for charts with isobars marked at 4 or 5 hPa intervals.

Weather associated with high pressure systems

High pressure systems, sometimes referred to as anticyclones, are often associated with settled weather and clear skies, but they can also bring low cloud, fog, humid conditions, or sometimes thundery conditions. In higher latitudes they may bring frost, especially in winter months and at night.

Centres of highs have light variable winds, and possibly fog. If the system is slow moving, conditions can be slow to change. Winds at the periphery, on the other hand, are often deceptively strong (40 or 50 per cent greater than winds at the same latitude, same isobar spacing but associated with a low). In trade wind areas they can add to the existing winds to produce reinforced trades that can last for a week or more.

---

[1] *Beware:* Charts from France, Germany and the Russian states are generally printed with isobars spaced at 5 hPa intervals. Those from Australia, the UK, Spain, Italy and New Zealand use 4 hPa intervals. Some 48- and 72-hour prog charts are printed with an 8 hPa interval.

Weather associated with low pressure systems

Low pressure systems, sometimes called cyclonic systems or depressions, have diameters ranging from between 100 miles to 2000 miles and can include a considerable variety of weather conditions. Also, because they move at speeds of anything from zero to 50 knots, over a few hours, a stationary observer may see the weather changing from clear to overcast with rain, clearing, then more rain and more wind. The actual sequence of events is important as it can be used to correlate the progress of a depression on the weather chart with the weather you actually experience. By noting cloud changes, onset of rain, wind shift, changes in pressure, temperature etc, it is possible to gain an accurate idea of your position in relation to fronts and the depression as a whole.

Fig 4.4 shows an idealised cross-section of a depression in the mid latitudes. Such systems tend to move from west to east and a fixed observer in the path of the two fronts would notice the following sequence of events:

1   First signs; **the head** The sky is gradually invaded from the west by cirrus and possibly thin altostratus cloud. Cirrus are wispy high-level clouds, composed mainly of ice, and sometimes hooked or comma shaped if there has been a change of wind direction. At this stage visibility is likely to be quite good ('too good to last'). Cirrus clouds could indicate the first intrusion of high-level moist warmer air from a coming warm front but, taken alone, are not conclusive. Other signs are a falling barometer, thickening cloud and a wind that slowly backs[1] (northern hemisphere) or veers (southern hemisphere).

2   **The body** The sky becomes darker and more overcast with a continuous layer of altostratus or nimbostratus. As the warm front approaches, rain sets in and may continue for several hours. The wind veers (northern hemisphere) or backs (southern hemisphere) to a more westerly direction and the barometer levels out.

3   **The warm sector** The air temperature begins to rise and rain eases to a light drizzle or stops. The barometer remains steady. The sky may remain covered with low stratocumulus or perhaps clear for a while before it becomes dark with a huge build-up of cumulus congestus and cumulonimbus as the cold front approaches.

4   **The cold front** brings the strongest winds, heavy rain and squalls. As it passes, the wind veers to the north-west (northern hemisphere) or

---

[1]Changes in wind direction are said to be backing or veering if they shift in an anticlockwise or clockwise direction.

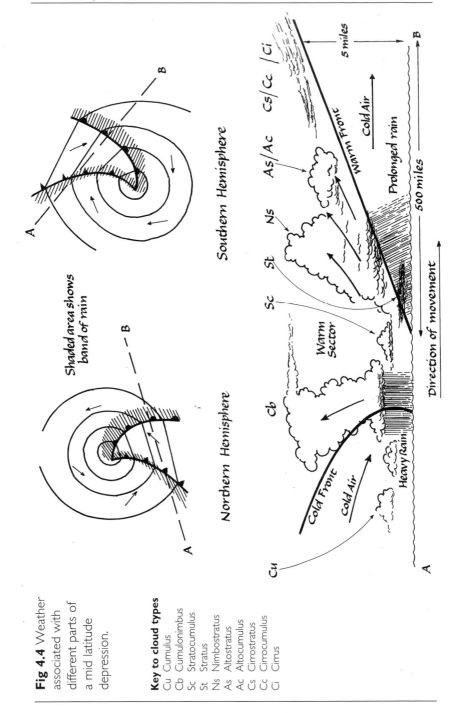

Fig 4.4 Weather associated with different parts of a mid latitude depression.

Key to cloud types
Cu Cumulus
Cb Cumulonimbus
Sc Stratocumulus
St Stratus
Ns Nimbostratus
As Altostratus
Ac Altocumulus
Cs Cirrostratus
Cc Cirrocumulus
Ci Cirrus

Shaded area shows band of rain

Southern Hemisphere

Northern Hemisphere

Cs/Cc /Ci

As/Ac

Ns

St

Sc

Cb

Cu

Warm Front

Cold Air

Prolonged rain

5 miles

500 miles

Direction of movement

Warm Sector

Cold Front

Cold Air

Heavy Rain

backs to the south-west (southern hemisphere). The temperature falls, but the barometer starts to rise.

5 **The wake** The sky begins to clear, but rain and isolated patchy squalls may persist. Winds are likely to be cold and strong but, unless a further depression is on its way, the barometer rises more quickly and larger cloud patches dissipate.

There are many variations on this theme depending on whether the depression has passed to the north or south of your location and the extent to which it has developed.

Fig 4.5 shows the cloud anatomy of a depression, with head, body, warm sector and wake areas. This is a slightly later development of that shown in Fig 4.4 and shows partial merging of the two fronts to form a single occluded front. This occurs because the cold front usually moves a little faster than the warm and eventually catches up. Occluded fronts share some characteristics of their component fronts with huge amounts of cloud, blustery conditions and winds veering (or backing) in the northern (or southern) hemispheres.

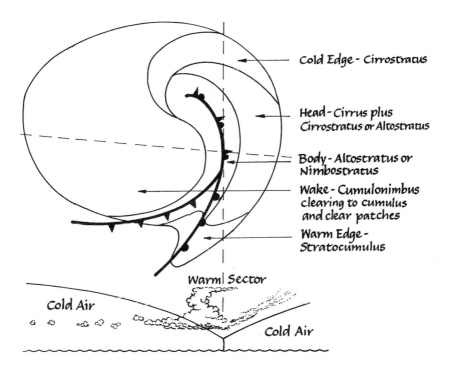

Fig 4.5 Cloud anatomy of a depression and occluded front.

Cloud type	Description	Appearance on satellite imagery
Cirrus (Ci)	High level wispy clouds often called 'mare's tails'.	Consisting of ice, it shows up well on IR images, but is usually too thin to appear on VIS.
Cirrocumulus (Cc)	Small white flakes like fish scales (mackerel sky) without shadows, and each about half the width of a finger held at arm's length. Sometimes occurring in banks.	Variable.
Cirrostratus (Cs)	High thin white cloud forming a veil across the sky. Occasionally only seen by the wide halos it produces around the sun or moon. (Cs is outflow cloud for Cb – see below.)	Consisting of ice, it's best on IR, but depending on density can show up on VIS. Streams ahead of active areas of Cb.
Altocumulus (Ac)	Mid-level clouds like cotton balls, but may have some grey content. Similar in structure to Cc but bigger and thicker – may be fused together and can cover the whole sky.	Light grey on IR, but good reflectors of light, so brighter on VIS.
Altostratus (As)	A mid-level veil-like cloud, thicker and lower than Cs. Greyish in colour and may cover the whole sky. Does not completely block the sun.	Variable.
Stratocumulus (Sc)	Low-level puffy clouds appearing in patches or banks. Grey with darker patches.	Variable.
Nimbostratus (Ns)	Thick, grey, low or mid-level rain clouds. Darkens the sky.	Dark grey on IR and VIS.
Stratus (St)	Grey low-level uniform fog-like cloud. May contain rain. Forms banks at sea.	Grey on IR, but bright on VIS.
Cumulonimbus (Cb)	The classic anvil-shaped thunder cloud with sharp edges.	A high-density cold structure. Shows up well on VIS and IR.
Cumulus (Cu)	Tall dense billowy clouds with flat bases. Often aligned in 'streets' with or across the direction of the wind.	White on VIS and grey on IR.

Table 4.2   Cloud types and their appearance from space.

Ridges and troughs

Ridges and troughs are shown on isobar charts in the same way as similar features are shown on contoured land maps. They are not always easy to pick out, but look for groups of isobars that curve together, then, bearing in mind that 'normal' pressure is 1013 hPa, check to see if pressures ridge up above this value or form a valley below it. Ridges show many of the characteristics of anticyclones (see page 58). Though they may not be marked as such, troughs often behave like cold frontal systems, bringing squalls and rain.

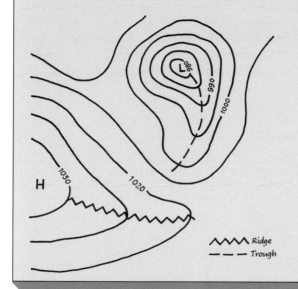

ᴧᴧᴧᴧ Ridge
— — — Trough

Fig 4.6
Ridges and troughs.

Upper weather influences

It is easy to forget that weather is a 3-dimensional phenomenon but, as was mentioned in Chapter 3, upper-level conditions can be a significant influence on the development of surface weather. In predicting weather up to a day ahead, a reasonably good forecast can be made simply by assuming that existing fronts, lows, highs etc will continue to progress with the same speed and direction they have had for the past 24 hours. For times further ahead, this simple procedure becomes less reliable, but clues from upper-level prognosis charts can be used to refine the projection to 36 or 72 hours.

Wind speeds greatly increase with altitude (logarithmically up to 600 metres). Like a twig that follows eddies in a stream, surface lows tend to

63

Omega blocking

This phenomenon occurs when jet stream winds assume the shape of a Greek letter omega (Ω) curving around a well-established surface high. It has the effect of blocking the normal west to east passage of surface features, forcing them to circulate around the high. In some cases it can cover a large area and remain in place for several weeks. As a result, regions within the path of the lows endure a stream of gales and rain, whereas those located under the high enjoy a similar period of settled conditions.

An example of such a situation occurs in the Atlantic when the stationary high normally located over the Azores comes far enough north in summer to bring settled weather over southern England and northern France. On the other hand, should the Azores high stay to the south, these countries are swept by a procession of depressions.

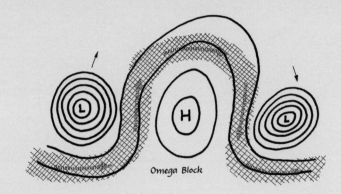

Omega Block

Fig 4.7 The Omega block.

follow the direction of the higher-level winds. In particular, winds at the 500 hPa level are a steering field for surface weather, and can be used to help predict future movements of lows and their associated fronts. As a general rule, said to be the result of many years of empirical observation, surface lows can be expected to move at half to one-third the speed of 500 hPa winds and in a path more or less parallel to the 5640 contour, though not necessarily immediately beneath it. Just how much cloud or rain is associated with a surface front also depends on upper air movements. It is not uncommon to find an upper-level trough lying above a surface anticyclone causing thundery rain.

Locating the jet stream on satellite images

Although not always easily spotted, the jet stream axis (ie path with the highest speed winds) is often indicated by a well-defined edge of cirrus cloud on its polar side. Known as *baroclinic zone cirrus*, this band of cloud lies about one degree distant from the stream and is more prominent on the eastern sides of curves. Occasionally it may be seen passing over lower-level cloud or, if there is insufficient water vapour, may be thinned and appear as streaks of cirrus.

Low pressure developments from space

Both IR and VIS images can provide useful clues to the location of developing low pressure and potential storm systems that usually follow a recognisable sequence of events. These begin with the formation of a *baroclinic leaf* cloud and proceed as in the steps shown in Fig 4.8. However, keep in mind that the simple existence of a *baroclinic leaf* is, in itself, no assurance that the development will continue.

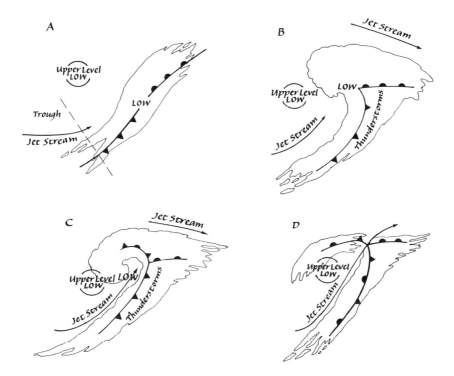

Fig 4.8 Diagrams A–D show the development stages in leaf to comma cyclogenesis (northern hemisphere).

Leaf cloud stage
These are usually formed on the eastern side of an upper level trough and appear as an elongated leaf shape. The side nearest the Pole has a distinctive border often curved in the form of a flattish S. On the western equatorial edge, the S is indented by jet stream winds pushing into the cloud formation.

Open comma stage
Air streams circulating around the point of maximum spin (ie vorticity) cause the leaf to develop into a comma shape with a ragged appearance to its eastern border. Where jet stream winds are crossing the cloud, the beginnings of a *dry slot* may appear. As the comma forms, surface pressures are usually falling. Comma clouds can appear in a range of sizes from small, localised thunderstorms to large cyclonic systems, and they can develop very quickly.

Occluded stage
Clouds spiral around the centre as the cold front catches up with the warm front to form an occluded front. Jet stream winds become cut off from the general circulation and the system loses its ability to deepen.

Shearing stage
Upper level winds cut the comma head from the tail, clouds become less organised, and the system continues to weaken.

Waves and swell
Whereas wind is the most important aspect of the weather for sailing boats, for power boats, waves and swell are the main concern. Fig 4.9 illustrates some basic terms used to describe all types of waves.

Because any group of waves will contain some that are larger than others, heights shown on fax chart forecasts (eg Fig 3.6) are given as significant wave heights. This is equivalent to the average height of the highest third of all waves. For those mariners chiefly interested in larger waves, the statistics of random waves predict that:

- One wave in 1175 is likely to be 1.9 times the significant height.
- One wave in 300 000 is likely to be 2.5 times the significant height.

Of course, these are only probabilities, and the bad luck of encountering 2 waves in succession, both 3 times the significant height, cannot be eliminated.

Sunglint
Even the largest of waves are far too small to appear on satellite images, though there are other clues that can be used to provide an indication of

the sea state. In VIS images, seas and lakes are normally very dark, almost black. However, on areas where they reflect sunlight towards the satellite, a localised bright region can appear, known as *sunglint*. The sun is best reflected when the sea is smooth. With stronger winds the reflected area appears larger and greyer, and as wind speeds increase still further it becomes darker and may disappear altogether.

Factors affecting the development of waves

Where waves are being generated, the most important factors that influence their size are the wind speed, the length of time it has been blowing and the fetch (ie straight line distance it has travelled). These effects are shown in the two graphs (Figs 4.10 and 4.11) which can be used as a basis for estimating wave heights from surface analysis or prognosis charts.

In practice, there are several other factors that influence wave development. Occasionally in gale force and higher winds, if it is raining heavily, seas are surprisingly smooth. Another important factor is the set and drift of surface currents which, if against the wind, creates steeper waves with a greater tendency to break. Other less important influences are:

● The width of the body of water (not a factor in open oceans).
● Air and water temperature difference.

At sea, it is common to find short-length waves superimposed upon much longer ones. Both are caused by the effects of wind blowing over water and, though there is no clear division, longer-period underlying waves are separately known as swell. Swells are simply waves that have travelled a long distance from their area of origin. As such, they may follow a different direction from wind and waves currently being formed. In mid ocean, swells can occur with heights of many metres and wave lengths of a mile or more.

Unless they have breaking crests, the steady rate of rise and fall as the ship passes over a swell is not necessarily uncomfortable or dangerous. With a low sun, it is a strange experience to watch the shadow of surrounding seas pass over the ship as it descends into the troughs. With the horizon only visible from wave peaks, in such conditions large ships may not be seen until they are dangerously close and astronavigation requires particular care as the true horizon is only visible from swell peaks.

Some regions are particularly prone to swells. In *Heavy Weather Sailing* by K Adlard Coles, the author recounts an incident where he encountered an exceptional swell while sailing off the Brittany coast in north-west France. The weather was bright and sunny with winds of only 4 to 6 knots (Beaufort force 2). The journey was hazardous because large seas made it difficult to spot navigational marks and, with so little wind and so much rolling, progress under sail was almost impossible. This is a

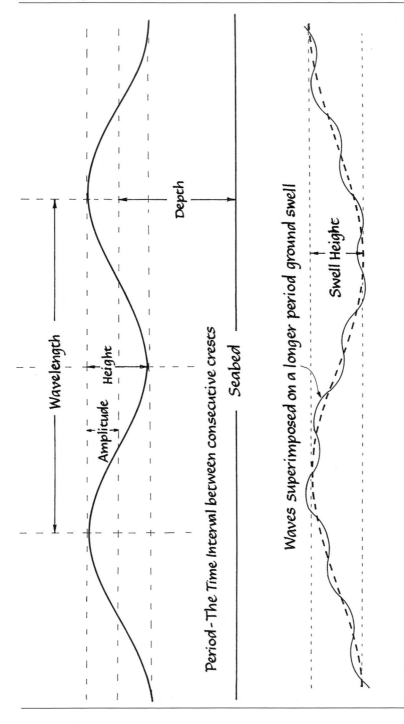

Fig 4.9 Terms used to describe waves.

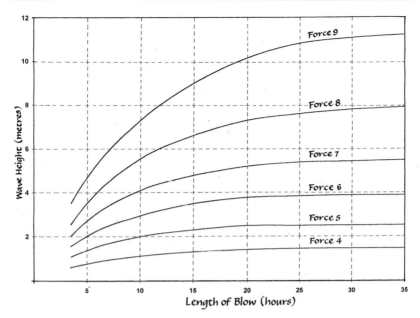

Fig. 4.10 Graph of wave height for the duration of a blow.

region where waves from the deep Atlantic pass over a relatively shallow continental shelf before meeting a very rugged coastline with numerous outlying rocks. On this particular occasion, the conditions were attributed to an exceptionally deep depression almost a thousand miles away in the north Atlantic. The example serves to illustrate how local conditions can be affected by distant weather systems. In the tropics, abnormal swells are sometimes a first visible sign of an approaching cyclone.

Fog

Some areas are more prone to fog than others. North-eastern Canadian and US coasts, the Bering sea and Aleutian islands are particular examples, though in the tropics it almost never occurs. As it has been a contributory cause of so many transport accidents on land and sea, it is most useful to know when and where it is likely to occur and, once established, how soon it is likely to clear. The usual fax analysis and prognosis charts give only indirect clues, but are useful if supplemented with an awareness of the conditions that encourage its development and a few local observations.

A first requirement is for air that is falling below its dew point (see page 35), and to measure this you need a set of wet and dry bulb thermometers. These are a pair of identical thermometers with the bulb of

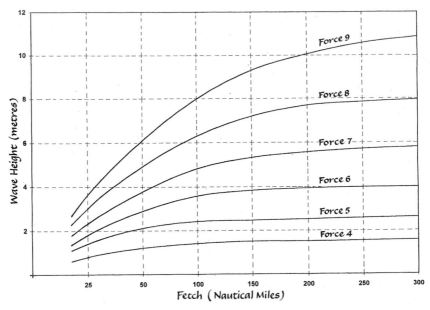

Fig. 4.11 Graph of wave height against fetch.

one covered by a damp wick. The uncovered thermometer measures free air temperature, but the temperature of the wet bulb is reduced by an amount corresponding to the rate of evaporation. The dew point is obtained by entering the difference between the two readings in the table in Appendix 5.

At sea, fog is likely to occur whenever the dew point rises above the sea surface temperature. To predict the likelihood of this occurring, the procedure is to take regular readings of sea temperature and dew point and plot the values as a pair of curves on the same graph. In this way, it is possible to monitor trends that would be likely to lead to the curves converging and thus the high probability of fog. Unlike sea water temperatures, dew-point readings can change quite rapidly, so to make the most of this method readings need to be taken at frequent intervals (10 minutes is the recommended maximum). In practice, automatic recording equipment is usually necessary.

Conditions likely to produce fog
Though the underlying mechanism is the same, several different types of fog have been identified and are associated with particular weather patterns. These are:

Advection or sea fog

This is formed when warm, moist winds blow over cold sea, so it is most likely where there are cold currents or in late spring when water temperatures are at their lowest. Once formed, sea fog tends to persist, but is dispersed if a change in weather brings rain. It is likely to thin or disperse completely if winds exceed 20 knots.

Frontal fog

This is found at the head of a warm or occluded front. It is caused by rain formed in the warm sector, which then falls through and raises the humidity of colder air below. Fog formed in this way is unlikely to be more than 50 miles in width, and disperses as the front passes over.

Sea smoke

This occurs when very cold, dry winds blow over warm seas. Steaming fog is caused by the wind cooling the warmer layer of damp air that is in immediate contact with the sea surface. This type of fog is more common in higher latitudes where it is known as arctic or frost smoke. It also occurs over inland lakes and rivers, often in the morning. It tends to thin and disperse later in the day if the sun warms the land and air temperatures rise.

Radiation fog

On clear nights, often when high-pressure weather conditions predominate, open skies allow land and air to lose heat by radiation. If the air is warm and saturated, fog is likely. Radiation fog is generally thickest in the early morning and, because its white upper surface tends to reflect heat, without a strong sun it can be slow to clear. Radiation fog mainly occurs over land, but can be found over seas up to 15 miles from coasts. Conditions for its formation are well defined and include:

- Moist air.
- Sky less than half covered by cloud.
- Winds between 3 and 9 knots. Too little wind causes the moisture to fall as dew, and with too much it is dispersed.

Smog (smoke-fog)

This is a peculiar type of land fog, formed in the usual way, but in air that contains small particles that encourage water condensation. These are frequently dust or smoke particles and often originate from some form of pollution. They make fog tend to form earlier and persist longer.

Chapter 5

Putting It All Together

'The isobars on the chart suggested a sunny 10 knots of wind, but outside it was raining and blowing a gale – could the chart be wrong?' It is tempting to think so, but unlikely. However, the habit of looking for relationships between chart features and the actual weather outside is an important step towards understanding weatherfax. In practice, situations regularly occur where, at first glance, the chart does not appear to back up the weather you are actually experiencing, but if you look a little deeper at the information on the chart and the way it is displayed, add your local knowledge of the area, then usually the differences can be explained.

Exceptions to the wind rules
The rules for predicting wind speed and direction given in Chapter 4 are a good general guide, but there are circumstances where they are less helpful or need to be modified, as the following examples show.

Terrain effects
Like water flowing around rocks in a river, wind speed and direction changes as it flows around obstacles such as a range of hills or a group of islands. Winds may be blocked by high mountains, or forced to rise over hills. They curve around headlands, accelerate down valleys, or form separate streams that recombine in another location. The effects are often very localised and too small to appear on weather charts.

Because atmospheric pressure reduces with altitude, a land barometer, under equivalent conditions to one at sea level, would read lower (1 hPa for every 10 metres altitude near sea level). However, on surface weather charts all isobar pressures are corrected to sea level values no matter whether they appear over mountains or oceans. When shown in this way their paths are changed, and it is no longer possible to assume that surface winds follow a similar direction.

In Fig 5.1 isobars show a gradient that slopes from high pressure in the north to low in the south but on crossing the island, prevailing winds are blocked by the mountains so local pressures increase. Because isobars are, by definition, lines of constant pressure, this effect causes them to

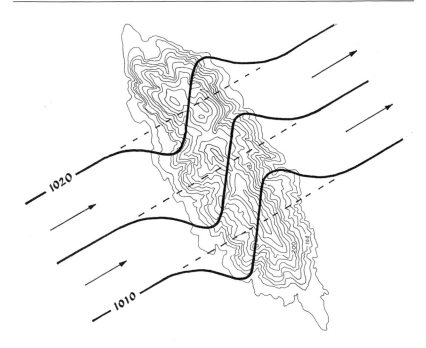

Fig 5.1 Isobar distortions caused by high ground (a southern hemisphere example).

divert from a straight line path towards the otherwise reduced pressure region in the south. On the leeward side of the slopes, the reverse process occurs and they swing back to resume their original route.

Land influences on winds at sea

Terrain effects also influence winds at sea and, if used wisely, are a guide to finding safe spots for anchoring, fishing, etc. In the Canary Islands, for example, the northeasterly trades have predominated for centuries – a fact that is confirmed by the numerous lopsided volcano craters, formed because erupting material is consistently blown away to the south-west. All the main islands have a good deal of high ground, and on south-facing coasts this can provide almost perfect shelter for boats as long as they keep close inshore. A local problem here is that water depths are usually too deep for anchoring except at Barrancos, which are a common feature of the area (Fig 5.2). These are bays with a central valley opening into a patch of shallower water. However, winds emerging are sometimes strong and gusty and may exceed the speed of winds over the islands as a whole. These effects are quite localised, and a few miles out to sea regular trade winds usually prevail.

Fig 5.2 Barranco del Parchel, Canary Islands.

Each island also makes a contribution to the pattern of winds around the archipelago as a whole. Winds funnelling between them are accelerated, often to gale force, when the average speed over the group as a whole is 25 knots or less. This produces patches of uncomfortably short, steep waves, sometimes in places well out of sight of land. In the Canaries, due to the consistency of the trades, their location is well known and is shown in Fig 5.3.

Rivers of wind
Similar effects are associated with blocks of high land in other parts of the world, but where winds are more variable in strength and direction, their location and severity changes accordingly.

Fig 5.4 shows a surface analysis of New Zealand broadcast on 24 March 1992. A cold front is passing to the south, but the area is dominated by a high pressure system centred in the Tasman Sea. Pressure gradients appear low over both islands and isobar spacings off the eastern tip of the North Island suggest winds in the region of 10 knots. However, knowing that East Cape is a mountainous region should suggest that things may not be entirely as they seem. Pressure gradients in this region were high but this only became clear when the intervening 1 hPa isobars were shown. On this particular occasion the vessel *Tangaroa*, well offshore from East Cape, experienced winds of 40 knots.

Another cape well known for this effect is Cape Finisterre in northern Spain. Typically, a high pressure system in the Atlantic, maybe with a ridge extending to France, will produce a northeasterly air flow over the ranges of Asturias and Galicia. The result is an offshore river of strong wind beyond the western coasts of Spain and northern Portugal.

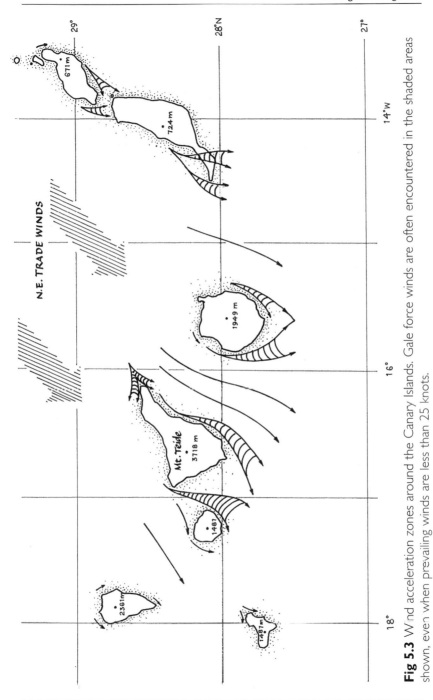

Fig 5.3 Wind acceleration zones around the Canary Islands. Gale force winds are often encountered in the shaded areas shown, even when prevailing winds are less than 25 knots.

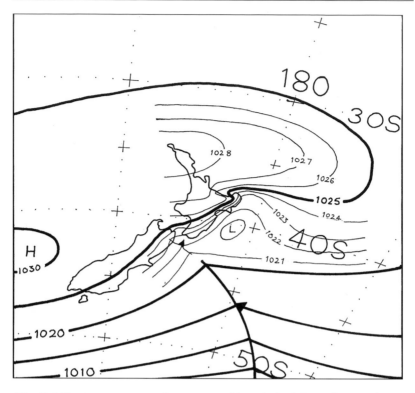

Fig 5.4 Steep pressure gradients, caused by winds deflected around mountains, are more clearly shown when isobars are drawn at 1 hPa instead of the usual 5 hPa interval.

Land and sea breezes

These winds are formed near coasts. They strengthen or weaken other winds affecting the area and are caused by differential land and sea temperatures. Whereas sea temperatures change little between day and night, land masses are inclined to warm up quickly during the day but lose heat quickly once the sun has set. During the day, warm air rising over land draws in cooler air from the sea, creating an onshore breeze or *sea breeze*. At night, the reverse situation occurs. With the land now cooler than the sea, *land breezes* blow offshore to replace air rising over the now warmer sea.

Sea breezes are usually stronger than land breezes and their effects are felt up to 10 miles out to sea. Both types tend to follow a regular diurnal cycle, dying away at dawn and dusk but reaching a peak near 3 am or 5 pm.

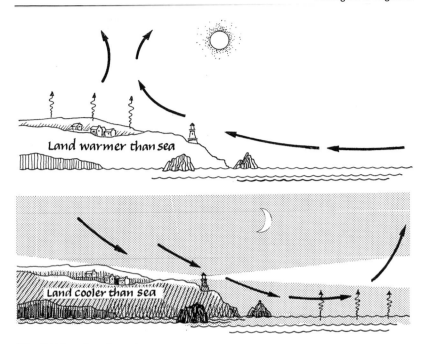

Fig 5.5 Land and sea breezes.

Katabatic winds

Like land breezes, these winds are also caused over cool land, but in this case the land is usually high and mountainous, often with deep valleys leading down to the coast (eg *rias* of northern Spain). A build-up of cold air remains trapped over high ground until eventually, like an avalanche, it cascades down mountain sides and valleys in strong gusts.

Katabatic winds can occur suddenly. Without warning, winds switch from a gentle breeze to gale force then back again, all within the space of a few hours. A special risk for boats sailing close inshore is the possibility of being blown down to a dangerous angle.

Limitations caused by shortages of raw data

In some areas, the tropics or thinly populated areas of the Pacific in particular, reporting stations are few and relatively widely spaced. Isobars are drawn as smoothed lines as a best-fit compromise between available data, but, as a result, small irregularities with local importance can be missed.

In the tropics, isobars are usually widely spaced, indicating light winds over the area. This may be deceptive; trade winds are often composed of separate bands of strong and weak winds. Taken as a whole, the average speed over an area may not be too great though it may contain localities with very much higher speeds.

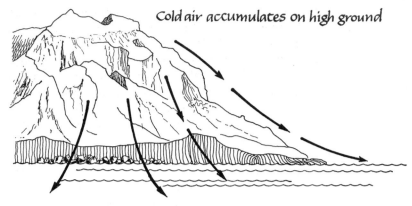

Cold air accumulates on high ground

Fig 5.6 Katabatic winds – caused by cold air descending from high mountains.

Improving fax maps with present weather observations

Timing

Weather charts are compiled from the best available information. When the chart is received, the data is several hours old. If a feature such as an approaching front is indicated, you have no reason to doubt its existence, but if it does not arrive as expected, the reason could be that its speed, direction or strength have changed since the chart was prepared. The difficulties are increased where there are fewer reporting stations. As a check on the approach of weather systems, it is useful to monitor your own weather conditions and listen to current reports from other stations.

Images from polar orbiters are also a good indicator of current weather conditions, and may allow you to spot baroclinic leaf cloud and the possible start of a developing low before its appearance on the current surface analysis chart. Keep in mind that the existence of leaf cloud is no certainty that severe weather will develop.

Fax pictures without a fax receiver

For years, the more painstaking boat skippers have prepared their own surface analysis charts by plotting observation data from reports from weather stations. Typically, these will include a barometric pressure and tendency, wind speed and direction, visibility and cloud cover. For western Europe these are conveniently appended to the end of the UK regional shipping forecast broadcast at 0048, 0555, 1355 and 1750 local time in a 5-minute slot between domestic broadcast programmes on 1500 metres (200 kHz). In other parts they are more usually broadcast by coastal radio stations on one of the marine band channels. Details of broadcast times, frequencies and services available can be found in the

Admiralty List of Radio Signals – Volume 3 or similar published lists of radio signals (see page 100). Report lists may be transmitted as a single side band voice signal or, more often, in code. Morse is still used by some stations, and if your skills at reading it still need developing, recording it and replaying at a slower speed can be helpful. Alternatively, computer software is available for decoding morse and some of the more popular types of telex.

By taking a map of the area marked with positions of the reporting stations and plotting barometric pressures, wind speeds and directions, it is possible to build a picture of where isobars are likely to lie and positions of significant weather features. In the UK, a basic map designed specifically for these broadcasts is available from the RYA.[1] In other areas, a photocopied section of a chart, enlarged or reduced as necessary, forms a good basis. You may prefer to copy off a number for routine use, or to use a transparent overlay and mark details with an erasable pen. At first the task seems dauntingly difficult, with precious little time to copy down a lot of information. With practice at listening to the broadcast format and by using symbols and abbreviations instead of longhand, it becomes easier and more relaxed.

Weather reports of this type form part of the data from which fax charts are prepared, but because processing it and drawing up the chart takes time, these reports are likely to contain more up-to-date information which is their main value. By plotting them directly on the fax chart, after advancing features by the appropriate time lapse, it is interesting to see what changes may have occurred.

If you have a fax receiver and are picking up regular charts, decoding and plotting station reports is a useful exercise but will probably not become a routine habit. However, if the weather contains critical features, perhaps fronts bringing wind shifts, rain or a system that could place you in a difficult situation, then current weather reports and the experience of using them is especially valuable.

Adding your own weather observations

Barometers

For sailors in the past, the barometer was their best and often only indicator of the coming weather. In fact, it still remains an almost essential piece of equipment to have aboard a boat; and fortunately, today's aneroid instruments have eliminated most of the disadvantages and inconvenience of earlier types.

Because much of the value of a barometer comes from recording rates of change of pressure rather than absolute values, a barograph is rather

[1]Royal Yachting Association, RYA House, Romsey Road, Eastleigh, Hampshire, SO5 4YA.

more useful than simple barometer. This could be the traditional type with clockwork driven paper chart that is marked by a pen attached to an aneroid sensing element or, alternatively, an all-electronic barograph. In this case output from a pressure transducer is converted to digital data that is periodically stored on some type of memory media. These instruments may be entirely self-contained with their own battery power supply and display, or integrated within a larger computer system where the results are downloaded for display and analysis when needed.

Siting

Aneroid barometers are sensitive to vibrations, so they should be mounted on a solid bulkhead, away from the engines or machinery. Though most include a mechanism that compensates for changes of temperature, a site away from direct sunlight or heat from cookers is advisable. Since their purpose is to indicate atmospheric pressure, clear access to the outside air is necessary. For this reason, in homes they are traditionally mounted in open hallways, rather than living rooms that may have tight-fitting doors and air conditioning or heating units. Aboard boats there is less choice, but it is as well to consider the effects of tight-fitting hatches and possibly a diesel engine sucking air from the accommodation.

Adjusting a barometer

Though rates of change of pressure are generally of more importance than actual values, it is often useful to be able to compare your current pressure with values appearing on the fax chart. To assist comparison between barometers at different surface altitudes, weather chart readings are corrected to mean sea level values which is the internationally agreed datum.

From time to time the barometer adjustment needs to be checked and, if necessary, reset and rechecked a few days later. For this purpose the reading is compared with that of another barometer of known accuracy in the same locality. Most airports and port meteorological offices are able to provide this information by phone, but the value used to reset the barometer should be corrected for altitude (see page 108).

Your barometer as an alarm clock

A barometer provides a useful check on how accurately features on the surface chart are keeping pace with actual events. Bob McDavitt of New Zealand MetService suggests using it as a kind of alarm clock:

'From the weather map decide on a target pressure associated with an event you are interested in (such as the passage of a front). Mark this target pressure on the face of your barometer, and then you can monitor the closeness of this event at a glance.'

Even with a good chart, time differences of around plus or minus a few hours are to be expected, so this is not the kind of alarm clock you would want to wake you up in the morning. None the less, it gives a useful indication as to how closely the chart is synchronised with reality.

Reporting wind and sea state

In these times of increasing reliance on electronics for environmental monitoring, wind strength and direction indicators have become a normal part of the cruising boat instrument inventory.

For some racing applications this has taken much of the guesswork out of assessing wind speeds and, used in conjunction with other performance indicators, gives a good guide to the behaviour of one sail configuration compared to another. Used in this way, it does not matter if the adjustment is not too good, just that readings are consistent.

The important point is that you are able to evaluate one day's performance against standards achieved on other occasions. If, on the other hand, the object of fitting wind instruments is to obtain readings that are meaningful when compared to meteorological forecasts, then the type of installation fitted to most boats at present leaves much to be desired.

A major difficulty is that winds measured aboard a boat are inevitably affected by the course of the boat through the water. *Apparent wind* experienced at sea is affected by the boat's own speed and direction, the speed and direction of any current and the speed and direction of the *true wind*. As an example, a boat anchored in a wind of 20 knots would be experiencing a fresh breeze, but a boat in the same conditions travelling at 9 knots directly into the wind would feel an on-deck wind speed of 29 knots or near gale. On the opposite course, the apparent wind would be reduced to 11 knots, giving only a moderate breeze.

Next there is the problem of where to site the anemometer. Meteorological observation stations use a standard height of 10 metres, which is somewhere close to the height of the average sailing boat mast, which would be the obvious place. Handheld anemometers at deck level are likely to be affected by turbulence around the boat's structure. Also, measurements at this level will be significantly lower than those at mast head height. Finally, there is the problem of calibration. With normal wear and tear and with time this may change, but most users will not have access to suitable standards against which it can be checked.

For reporting wind and sea conditions between boats, the simple low-technology solution to these difficulties is to use the Beaufort Scale. This is an indirect method that relies upon observing the effects of wind upon the sea condition. After a steady blow for an hour or so, the sea acquires an appearance that can be correlated with the strength of the wind that produced it (see Appendix 6). Recognising Beaufort sea states takes a little practice, but the method is unaffected by short duration gusts. A concise statement, such as 'North Westerly Force 4', is a meaningful report of wind speed, direction and sea state. The scale was first introduced by Admiral Beaufort in 1808, and is still widely used for meteorological reporting today.

Radio nets

These are radio meeting places, where boats in a particular area call in to exchange information of mutual interest. They can be found on both marine band and amateur radio frequencies. Some (such as the United Kingdom Maritime Net) have been in existence for 25 years or more while others last only a single season. No one is exactly sure how many are in regular operation but they are numbered in the hundreds. Table 5.1 lists a few of the longer-standing nets.

Net	Time UTC	Frequency MHz
Intermar (Germany)	0700	14.313
Pacific Maritime Net (US)	0300	14.313
South Africa Net	0630	14.316
	1130	
Tony's Net (New Zealand)	2100	14.315
Traveller's Net (Australia)	0200	14.116
United Kingdom Maritime Net	0800	14.303
	1800	14.303

Table 5.1 Amateur radio nets concerned with reporting weather.

For those interested in the weather, the main value in monitoring amateur nets is that they provide an alternative source of weather information. Weather condition reports from other boats can be plotted to provide a further check on the progress and intensity of approaching weather features.

Chapter 6

Storms, Hurricanes and Cyclones

I n a typical lifetime of sailing, the misfortune of a firsthand encounter with a severe tropical storm is, fortunately, not a regular event. However, if confronted by one – either at sea or in harbour – the chances are slim that you and your craft will survive unscathed. For this reason and because such storms are not restricted to the tropics, this chapter is devoted to the subjects of early warning and avoidance.

In the Caribbean and north-east Pacific, tropical revolving storms are known as hurricanes. In the north-west Pacific they are typhoons, and in the Indian Ocean and South Pacific they are cyclones. The names refer to the same type of weather system, though they occur at different times of the year. Fig 6.1 shows the principal areas and months of highest probability.

Development and evolution
Though the actual event that triggers the development of a true tropical revolving storm is uncertain, there are several preconditions without which they are not able to develop. Essentially these are:

- *Sea water surface temperature above 26°C.* This only occurs within the tropics.
- *Moist air.* Tropical revolving storms do not begin over land.
- *Spin.* In the northern hemisphere, winds in low pressure systems rotate in an anticlockwise direction. In the southern hemisphere, the direction is clockwise. Close to the Equator, the Earth's rotation imparts no spin and winds blow in straight lines. As a result, rotational storms never cross the Equator and are unheard of in latitudes of less than 5°.

A favourite spot for storm development to begin is the ITCZ (see pages 30–1). This is a location where thundery squalls are commonplace and are associated with rising warm, moist air and heavy build-up of cumulonimbus cloud. In summer months, sea surface water temperatures may

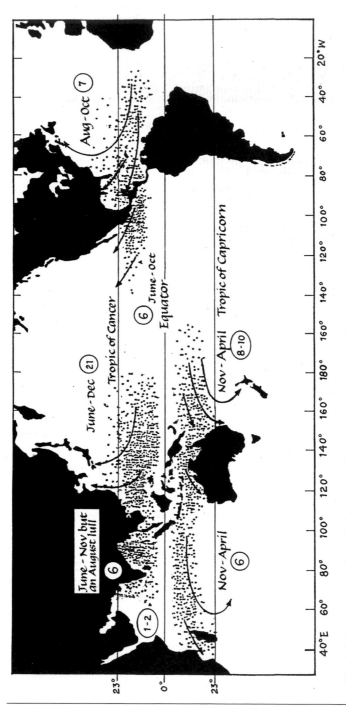

Fig 6.1 Occurrence of tropical cyclones. Dots represent points of origin and circled figures the average number per year (after Gedzelman, 1980).

The South Pacific storm of June 1994 was hardly felt on land, though it hit the New Zealand to Fiji/Tonga route at the peak of the sailing season. Six boats were abandoned, and one lost along with all crew members. Here, Robyn and Bill Forbes are hauled aboard HMNZS *Monowai* in winds of around 70 knots. (Crown copyright)

exceed the 26° threshold and, if the latitude is greater than 8°, converging winds may acquire sufficient angular momentum to form a rotational weather system. Here, the centre, or 'eye', forms an exit point for converging winds. Warm moist air is drawn in from a wide area and rises near the centre, forming a deep low pressure region. As the wet air ascends, it cools and expands, forming copious clouds and releasing heavy rain. At a higher level, the winds spiral outwards in an opposite direction and are distributed over a wide area.

Thermodynamically, a rotational storm is a type of heat engine. Rather like a steam engine that changes heat into mechanical energy, the storm engine converts heat from sea water into kinetic energy as wind. Truly enormous quantities of energy are transformed in the process and, though in human terms it all goes to waste, the mechanism provides a natural means of cooling ocean hot spots.

A tropical storm may take several days to achieve full strength and always passes through four developmental stages that are recorded on charts as follows (page 87):

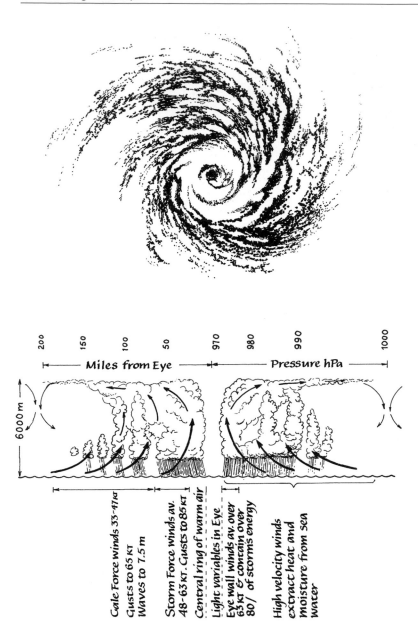

Fig 6.2 Anatomy of a tropical storm (southern hemisphere).

1 *Tropical disturbance:*   — Winds less than 28 knots.
2 *Tropical depression:*   — Winds 28–33 knots.
3 *Tropical storm:*   — Winds 34–63 knots.
4 *Hurricane:*   — Winds over 63 knots.

If one condition or another is not quite right, many storms just peter out or do not go on to develop very high winds. Only when average speeds exceed 33 knots (gale force) are they given an identifying name[1] which is used until the system dissipates, even if the wind speed later falls below this level.

According to traditional wisdom, tropical storms are formed at around 10° latitude, then move steadily in a westerly direction, before 'recurving' and heading towards higher latitudes with an easterly component to their course (see Fig 6.4). In the early stages of formation they may move at speeds of up to 10 knots or a little more in higher latitudes. After recurving, movements of 20–25 knots are typical, though they may reach 40 knots or more.

In recent years, it has become evident that very few tropical storms actually behave in such a predictable manner. They may speed up or slow

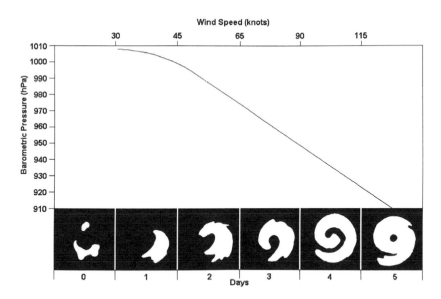

Fig 6.3 Tropical storm system development. Satellite view of cloud formation (northern hemisphere) as central pressure falls and wind speeds increase.

[1]Lists of names are kept by the Regional Cyclone Warning Centres. These are selected from lists compiled by the Tropical Cyclone Committee of the World Meteorological Organisation.

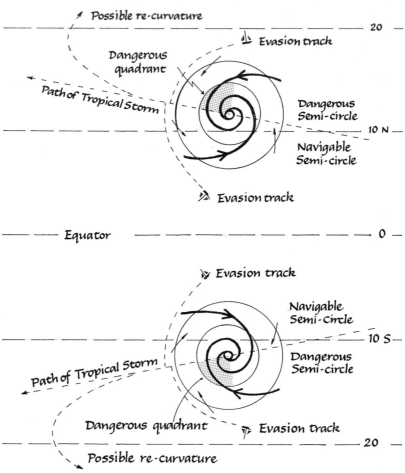

Fig 6.4 Classical tropical storm tracks and avoidance strategies.

down or remain stationary for a while. Some may move backwards or loop back and re-cross their previous track. In the south Pacific, more than 50 per cent of cyclones do not follow conventional routes. This is not to suggest that plotting tracks is of no value, but simply that predictions based on their past progress should be treated with circumspection.

Early warning signs
- Watch for isobars with a small kink towards the nearest Pole. Occasionally these form closed loops and the beginning of an active low pressure system.
- Low pressure systems with central pressures less than 1000 hPa.

- Rapidly falling barometer. A drop of 3 hPa in 3 hours, after allowing for diurnal variations and latitude (see page 107), should be taken as a serious indication that a storm may be imminent.
- Unusually heavy swell from an abnormal direction.
- An abnormally long, low ocean swell appearing from the direction of the storm centre.
- Watch for extensive cirrus cloud followed by altostratus, then broken cumulus or scud.
- Baroclinic leaf cloud on satellite imagery.

Note: A small but intense storm may give little warning of its approach.

Dangers from tropical storms
These include:

- Extremely heavy rain falling like 'sheets of continuous spray'. Extensive flooding and land slides. Bridges may be washed away.
- A heavy ocean swell with large breaking waves. Sea walls and harbour defences may be damaged; coral may be smashed and flung into the air. On the island of Niue, a hotel at the top of a 25 metre cliff was severely damaged by seas.
- A storm surge caused by high winds and reduced pressure may lead to an increase in sea level of 1 or 2 metres. If this coincides with the time of high water, boats may be washed ashore and low-lying land could be completely covered.
- Extensive destruction caused by strong winds – particularly those blowing through narrow straits and around headlands.
- Danger from flying debris, eg corrugated steel roofing sheets, parts of trees and buildings.
- After the storm, public health may be affected: sewage disposal disrupted, fresh water supplies polluted; domestic animals and food crops destroyed. Electricity supplies and communications may be disrupted.

Avoidance
For small boats, by far the best tactic is to make sure that you are not in a tropical storm area within the storm season. Storms are rare in latitudes of less than 10° and, for this reason, Trinidad and Venezuela are low-risk areas of the Caribbean. In the Pacific, islands in the Phoenix group or the Solomons are a possible storm season refuge. Latitudes higher than 35°–40° could also be considered, though no part is totally free from risk.

If you stay in an area where tropical storms are likely, it is important to have a range of strategies planned before any storm is forecast. In a small boat, putting to sea is only a possibility if you have time and the necessary speed to avoid the system or are able to reach a better protected or safer harbour.

Extra-tropical storms

Extremely destructive tropical rotational storms are usually no more than a few hundred miles in diameter. When they move out of the tropics, they encounter colder waters and usually dissipate. Some expand into a much larger system, still with strong winds but maintained by a different mechanism.

On meteorological charts they continue to be identified by their earlier name, but lose their circular symmetry and develop the characteristics of a typical temperate-latitude low pressure system. The central eye is no longer a region of warm clear air, but of violent convection and heavy rainfall. There are one or more frontal systems and the storm moves in an easterly direction making slightly more (southern hemisphere) or less (northern hemisphere) than a right angle with the occluded front. Highest winds are experienced ahead of the fronts; and though these may not be as great as those of the tropical system, they can still pack a hard punch.

Case studies

Every few years, the public interest is captured by some spectacularly dramatic storm at sea. A number of sailing boats are usually involved and the media cover stories of extreme personal grief, tragedy and heroism. In the following sections we look at three such storms, where a common factor was that all gave few signs that they would develop so rapidly and with such severity. For some boats that were already at sea, there was far too little time to take avoiding action or seek shelter.

The Perfect Storm (1991)

Unlike the other storms covered in this section, this one did not involve large numbers of amateur sailors, but instead gained its popular notoriety through the book of the same name by Sebastian Junger, and later the film. Originally dubbed the 'Halowe'en Storm', it was renamed by meteorologist Bob Case because of the set of circumstances that aided its development.

In October/November, cold conditions are well established throughout much of Canada. Cold winds begin to invade much of the mid-west USA, though waters of the Atlantic still hold some of their summer heat. On fax charts of 26 October 1991, events leading up to the storm were starting to become clear, with a region of high pressure established over Canada and a cold front over central states of the USA that was also moving eastwards. Out in the Atlantic, east of Bermuda, tropical cyclone Grace was heading north and had already produced large seas, but was now in its final dying stages.

Supported by an upper level trough, the cold front continued to move east, approaching Grace on 29 October and forming its own new low over the warmish waters of the Gulf Stream. With high pressure (1043 hPa) to the north-west bringing in cold air, and the upper level trough providing extraction, the low continued to deepen. On 30 October the central pressure bottomed at 972 hPa. With steep pressure gradients creating a squash zone between it and the band of high pressure to the west, sustained winds of 60 knots were recorded.

In common with extra-tropical storms in general, this one affected a very large area, though its end is a little unusual. On 1 November, it encountered warmer waters, and began to acquire the more compact characteristics of a tropical cyclone with sustained wind speeds of 65 knots. However, this stage was short-lived and it was never officially named.

Effects of the storm

Much of the eastern coast of the USA from Nova Scotia to New Jersey experienced exceptionally strong winds and high seas for several days. With tides of 1–2 metres above normal, much damage was caused by coastal flooding and heavy surf. Sea defences were destroyed, many boats sunk at their moorings, and hundreds of homes and businesses were lost.

At sea, wave heights were phenomenal. A NOAA buoy at about 41°N 67°W reported gusts up to 65 knots and wave heights of 11.9 metres, though there were many unsubstantiated reports of even more severe conditions. In the book and film entitled *The Perfect Storm*, an account is given of imaginary events leading up to the loss of the swordfishing boat the *Andrea Gail*, and also the abandonment of the 32-foot sloop, *Satori*, where a distress message was issued by the traumatised crew rather than the skipper. Contrary to the book and film versions, the skipper of *Satori*, Ray Leonard, hotly denies accusations of drunkenness and incompetence. As a holder of a US Coast Guard 100 ton Master's License and with many years of experience with small vessels, he claims that it was the Coast Guard that ordered him to abandon. The fact that *Satori* was later found to have survived the storm illustrates that it is possible for a well-found small boat to endure the most severe of weather. At the time of writing, Ray Leonard still lives aboard *Satori*.

'The 1994 Queen's Birthday Storm'

This South Pacific storm began forming on 1 June 1994 near Vanuatu, and is remembered for the effect it had on a large number of sailing boats sailing between New Zealand and Tonga. Many were taking part in the New Zealand Island Cruising Club's annual sail to the islands. Six boats were abandoned and their crews picked up by other vessels. One of these, *Quartermaster*, was lost, along with its crew of three.

This was not the cyclone season and the storm that developed was

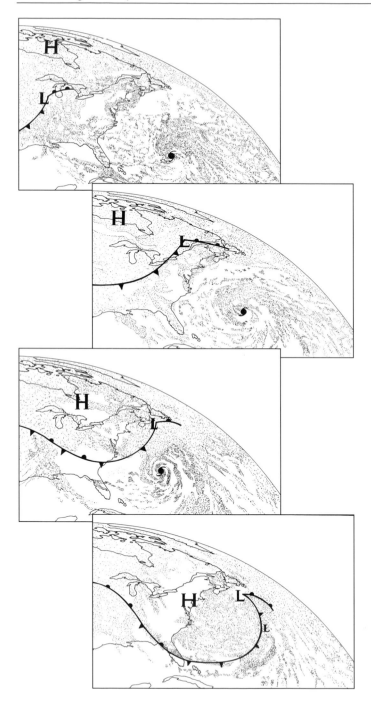

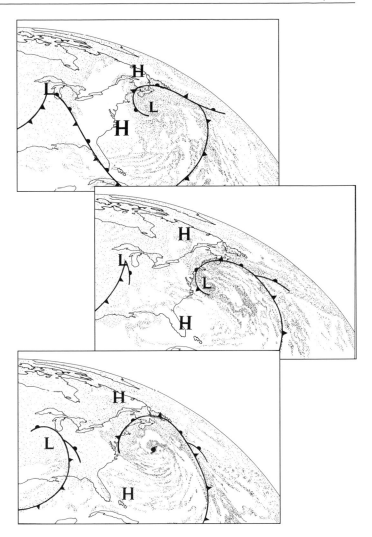

Fig 6.5 Formation of the 'Perfect Storm'.

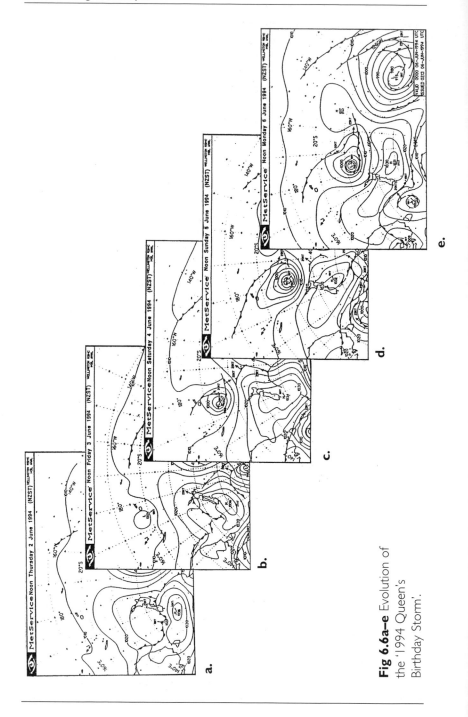

Fig 6.6a–e Evolution of the '1994 Queen's Birthday Storm'.

never officially named as such. In any case, it had no core of central warm air that is characteristic of cyclones, and hardly reached gale force while it was in the tropics. None the less, its effects in the sub-tropics were devastating. Fig 6.6a shows the mean sea level analysis on 2 June, with a slight kink in the 1010 hPa isobar near Vanuatu as the only indicator of what was about to take place.

As with the Perfect Storm, all the ingredients for rapid intensification – ie warm and cold air flows plus an upper level extraction – were present. Of particular significance is the established area of high pressure over New Zealand that brought in a supply of cold air from the Antarctic. Cold air does not mix easily with warm air, and the effect of this inflow was to force the existing warm air upwards. Not shown on the surface analysis is the presence of upper-level winds that, in this case, were able to withdraw rising air faster than the incoming cold stream was able to displace it. Barometric pressure at the surface was reduced still further as the system increased in size.

This phenomenon of cold air being drawn into a deepening low pressure system is sometimes referred to as a meteorological 'bomb'. The practical consequence is that weather conditions can rapidly switch from fair to foul. In this case, barometer readings gave little warning of what was in store and the storm moved too fast for the average sailing boat to outpace it. The 'bomb' is also commonplace in the north Atlantic, and adds to the severity of low pressure systems crossing north-western Europe. The 1979 Fastnet Race storm was another notable example.

Tactics
Boats caught in the '1994 Queen's Birthday Storm' experienced very large seas and winds of over 50 knots for up to four days. They were fortunate in having plenty of sea room in all directions and used a variety of tactics to cope with the conditions. Those to the west of its track were effectively in the 'navigable semicircle' and many ran with the wind astern, using drogues to slow the boat speed and reduce the risk of being pooped.

Those to the east of the storm track found themselves in the 'dangerous semicircle'. The best route out of the storm was to sail north and east, though wind and sea conditions made this impossible. In these conditions, most boats chose to heave-to under various combinations of storm sail, a technique that many found effective and relatively comfortable. A few sustained damage from occasional, exceptionally large, breaking waves, crashing against the hull.

The 1998 Sydney–Hobart Race
On 26 December 1998, 115 boats sailed from Sydney on the 54th Sydney to Hobart Yacht Race. The 630-mile course takes boats across the Bass

Straits that lie between the Australian states of Victoria and Tasmania. Like several other similar straits around the world, it's a region well known for rough conditions, though 1998 was to prove exceptional as only 44 boats reached their destination. Several boats were abandoned, and six people died.

The fax chart available to race skippers on departure shows key features in events that were about to unfold. There is a low pressure trough to the east of Tasmania ahead of a cold front in the Australian Bight. A deepening low is moving south between Australia and New Zealand and a high pressure centre to the south-east of New Zealand extends a ridge across the Tasman to New South Wales. Not shown on this type of chart is the fact that this system was blocked and slow moving.

At this stage, forecasters were finding it difficult to say exactly how these systems would develop since the various atmospheric computer models were predicting different outcomes. For race boats, initial indications were for a fast passage with north-easterly winds of 25 to 35 knots, but just an hour into the race, the Australian Bureau of Meteorology (BOM) issued a storm warning for the Straits with a change of west to south-westerly winds.

By the morning of 27 December, the situation became much clearer. The cold front from the Australian Bight had caught up, and merged with the trough to form a low pressure area over the eastern Bass Straits. Aided by an upper level trough, it deepened as it moved east-north-easterly through the Straits. Fig 6.7 shows how the system evolved. Note how warm and cold air flows, important ingredients for storm development, were provided by flows around the high pressure systems to the east and west of Australia.

The official enquiry – weather-related findings

Throughout the 27th, on the *Esso Kingfish* B platform located in the Bass Straits, mean wind speeds of 35 to 48 knots with gusts to 64 knots were recorded. Also in the Bass Straits, the Wilson's Promontory lighthouse, at 9 am on the same day, reported winds up to 79 knots. However, it was concluded that since its recording instrument was 100 metres above sea level instead of the standard 10 metres, this figure should be reduced by some 20 to 25 knots. It has also been suggested that terrain effects may have further disturbed the readings, but these would also be felt by boats at sea.

At the height of the storm many boats reported wave heights and wind speeds considerably higher than those recorded by the forecasters. In the case of wind speeds it is unclear which ones were inferred from sea conditions, and which ones were taken from instrument readings. Of those that used instruments, questions remain about where the sensors were located, the accuracy of their calibration, and if the speeds reported were of true or relative wind, 10-minute averages, or gust peaks.

From the report of the New South Wales coroner, it is clear that quite a number of race organisers, and even fairly experienced boat crews, did not appreciate the meaning of the terms 'mean wind speed' and 'significant wave height', as used in wind and sea state forecasts (see page 81). Many were surprised to learn that forecast figures were not for maximum wind and sea conditions.

Conclusions about these storms

As severe as all of these storms undoubtedly were, none was unprecedented, but they grabbed the headlines simply through the deaths and destruction caused. In the case of the Queen's Birthday Storm, for example, the fax charts of Fig 6.6 actually show another depression of similar ferocity, but since no one is known to have been affected, it passed unnoticed. The message here is to avoid being caught in the wrong place at the wrong time. Obtain the weather information that's appropriate for your needs, make a critical analysis, and decide for yourself what developments are likely and their effects upon your plans.

This is, in effect, the message of this book. However, I do not pretend that avoiding a serious storm is always easy or even possible. Along with my wife and daughter, I too was caught up in the Queen's Birthday Storm, though not as part of the rally. Heavy weather preparation and tactics are fundamental skills for which there can be no substitute, but how much better it is to spend time studying and preparing for storms than it is to actually have to fight them.

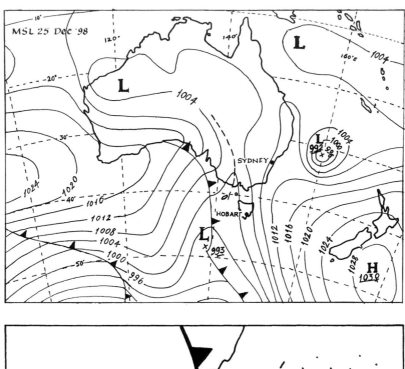

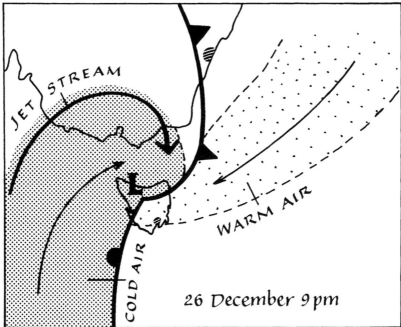

Fig 6.7 Surface pressures on 26 December 1998 at the start of the Sydney–Hobart Race, and the development and progress of the storm system.

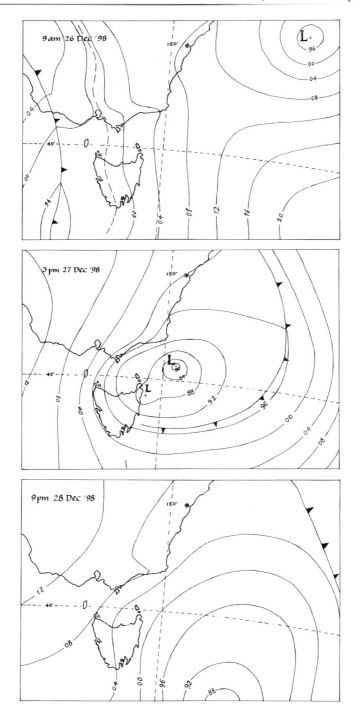

Appendix 1

Hf weatherfax station browser

Location	Station	Call Sign	Frequency (MHz)*
Alaska	Kodiac (USCG)	NOJ	20.54 4.298 8.459 12.4125
Antarctic	Casey Meteo	VLM	7.470
Argentina	Puerto Belgrano	LSR LSR	5.705 12.672
Australia	Charleville	VMC VMC VMC VMC VMC	2.628 5.100 11.030 13.920 20.469
	Wiluna	VMW VMW VMW VMW VMW	2.628 5.100 11.030 13.920 20.469
Canada	Iqaluit, NWT	VFF	3.2511 7.7081
	Resolut, NWT	VFR	3.2511 7.7081
	Halifax,	CFH	0.1225 4.271 6.4964 10.536 13.510 2.754

* The relationship between frequency (f) in MHz and wavelength (λ) in metres is: $f = 300\backslash\lambda$ or $\lambda = 300/f$.

Location	Station	Call Sign	Frequency (MHz)*
China	Bejing Meteo	BAF 6	5.5269
		BAF 36	8.1219
		BAF 4	10.1169
		BAF 8	14.3669
		BAF 9	16.0259
		BAF 33	18.2369
Denmark	Copenhagen Skamlebaek	OXT(1)	5.850
			9.360
			13.855
			17.510
Germany	Hamburg/ Pinnenberg	DDH 3	3.855
		DDK 3	7.880
		DDK 6	13.8825
Greece	Athens	SVJ 4	4.481
			8.105
India	Delhi Meteo	ATP 57	7.4049
		ATV 65	14.842
Italy	Rome Meteo	IMB 51	4.7775
		IMB 55	8.1466
		IMB 56	13.5974
Japan	Tokyo Meteo	JMH	3.6225
		JMH 2	7.305
		JMH 4	13.597
		JMH 5	18.220
Korea	Seoul Meteo	HLL 2	5.385
			5.8575
			7.4335
			9.165
			13.570

Location	Station	Call Sign	Frequency (MHz)*
New Zealand	Wellington	ZKLF	3.2474 5.807 9.459 13.5505 16.3401
United Kingdom	Northwood	GYA	3.2895 2.6185 8.040 11.0865
USA	Boston	NMF	4.235 6.3405 9.110 12.750
	Honolulu	KVM70	9.9825 11.090 16.135 23.3315
	New Orleans	NMG	4.3179 8.5039 17.1464
	Point Reyes California	NMC	4.346 8.682 12.5905 17.1512 22.527

The station list included here is but a small sample of the many that are available. Because frequencies and operating schedules are often changed readers are recommended to refer to current lists that can be obtained from: www.nws.noaa.gov/om/marine/radiofax.htm.

Characteristics of fax transmissions

Type	Drum speed	IOC	Modulation frequency (Hz)	Band width (Hz)
Amateur	120	288	829	2458
Weather	120	288	905	2610
	60	288	452	1704
	60	576	905	2610
	90	288	679	2158
	90	576	1357	3514
	120	288	905	2610
Re-broadcast satellite images	120	576	1810	4420
	240	288	1810	4420
	240	576	3619	8038

Index of co-operation (IOC) is defined by the formula:

$$IOC = F \times L/\pi$$

Where L = length of the scan line
F = scanning density
(ie number of lines per unit depth)

Appendix 3

Map projections

Most weatherfax charts cover a large area of the Earth's surface, but since the Earth is spherical and fax charts are flat, it is inevitable that charted features are to some extent distorted. The geometrical technique used to transfer shapes of features from a curved surface to a flat sheet is known as the map projection. Depending on the choice of projection, scaled directions, distances, areas and angular relationships may be changed. Though an infinite number of different projections are possible, in practice only a dozen or so are in regular use. Fig A.1 shows some simple examples.

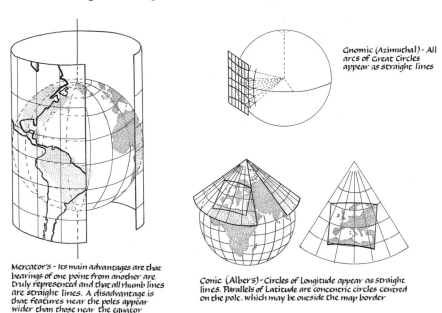

Gnomic (Azimuthal) - All arcs of Great Circles appear as straight lines

Mercator's - Its main advantages are that bearings of one point from another are truly represented and that all rhumb lines are straight lines. A disadvantage is that features near the poles appear wider than those near the equator

Conic (Alber's) - Circles of Longitude appear as straight lines. Parallels of Latitude are concentric circles centred on the pole, which may be outside the map border

Fig A1 Examples of map projections.

Weatherfax charts are produced in a variety of projections, depending on their purpose and area covered. A Mercator chart, for example, is often used to cover latitudes close to the tropics. For higher latitudes a conic projection centred on the Pole gives a truer representation of land areas, but directions are differently represented across the chart.

Symbols used on weather charts

Cloud cover (shaded area within station circle)

○	Clear sky
◐	One okta
◕	Two oktas
◑	Three oktas
◑	Four oktas
◒	Five oktas
◕	Six oktas
◑	Seven oktas
●	Eight oktas
⊗	Sky obscured

H	Centre of high pressure (anticyclone)
L	Centre of low pressure
➔ 20 KT	Direction and speed (KTS) of high or low pressure
▲▲▲	Cold front
●●●	Warm front
●▲●▲	Occluded front
▲●▲●	Stationary front
⌒⌒	Convergence line
▬ ▬ ▬	Trough line
—92—	Isobar line. Number denotes pressure in hPs. Leading 9 or 10 omitted.

Clouds

⟶	**High level:**	Cirrus
∠		Cirrostratus
⌇		Cirrocumulus
⫽	**Medium level:**	Altostratus
⌣		Altocumulus
⏀	**Low level:**	Cumulus
---		Stratus
⊃○⊂		Stratocumulus
◿		Nimbostratus
⏀		Cumulonimbus

Wind arrows showing strength and direction

⊙	Calm
\＿	Half barb = 5 kts
\＼	Full barb = 10 kts
\\⦀＿	25 kts
\\⦀⦀＿	45 kts
▲＿	50 kts
▲\\＿	65 kts
▲▲＿	100 kts
⟋⦀	35 kts from NE
⟍○	15 kts from W and includes reporting station circle
▬⫿⫽⫽⫽ FL370 ▬ 130 kts at 37,000′	Jet stream with speed & flight level height of max wind

Position of symbols around station circle

⟶	High level
⌣	Medium level
○	Station circle
⊃○⊂	Low level

Storms

 Tropical disturbance

⊗ Tropical depression
(cyclonic winds 33 kts or less)

Ϙ Tropical storm
(cyclonic winds 34–63 kts)

 Hurricane/Cyclone/Typhoon
(cyclonic winds 64 kts or more)

Temperature

Temperature in degrees Celsius are indicated by figures in circles. Air temperatures are negative unless prefixed by a plus (+) sign

(+7) + 7°C

(21) – 21°C

Precipitation

, Drizzle

▽ Shower

● Rain

✳ Snow

△ Hail

ʀ Thunderstorm

These may be combined eg:

 Rain showers

 Heavy rain

 Sleet

Their position around the station circle indicates:

Pressure

╲ Falling

╱ Rising then steady

⌐ Steady then falling

∧ Rising then falling

Position of symbols on the station circle

Actual pressure
Pressure trend
Change of pressure in last 3 hours in tenths of hPa

Visibility

47 Actual visibility in tenths of km

═ Mist

≡ Fog

Position on station circle

present weather / past weather

Turbulence

⌃ Moderate turbulence in cloud

⌃ Severe turbulence in cloud or clear air

Ψ Moderate aircraft icing

Ψ Severe aircraft icing

Others

390 Triple digit numbers give tropopause height in fl 39,000 feet

Boundary of area containing significant weather, eg thunderstorms

Area of gales, storms or fog

 Area and movement of high wind force 8 or more

 Area of heavy swell – 4m or more. Height and direction of swell is shown within broken line. Direction and speed of movement arrow is shown outside

107

Appendix 5

Tables and conversions

Sea level corrections to be applied to barometers at various heights and air temperatures

Height metres	Temperature (°C)										
	-10	-5	0	5	10	15	20	25	30	35	40
5	0.6	0.6	0.6	0.6	0.6	0.6	0.6	0.6	0.6	0.6	0.5
10	1.3	1.3	1.3	1.2	1.2	1.2	1.2	1.1	1.1	1.1	1.1
15	1.9	1.9	1.9	1.8	1.8	1.8	1.7	1.7	1.7	1.7	1.6
20	2.6	2.5	2.5	2.5	2.4	2.4	2.3	2.3	2.3	2.2	2.2
25	3.2	3.2	3.1	3.1	3.0	3.0	2.9	2.9	2.8	2.8	2.7
30	3.9	3.8	3.8	3.7	3.6	3.6	3.5	3.4	3.4	3.3	3.3
35	4.5	4.5	4.4	4.3	4.2	4.2	4.1	4.0	3.9	3.9	3.8
40	5.2	5.1	5.0	4.9	4.8	4.7	4.7	4.6	4.5	4.4	4.4
45	5.9	5.7	5.6	5.5	5.4	5.3	5.3	5.2	5.1	5.0	4.9
50	6.5	6.4	6.3	6.2	6.0	5.9	5.8	5.7	5.6	5.6	5.5

Diurnal variations of barometric pressure for 0° to 20°

Local time (hours)	Lat N or S 0°–10° hPa	Lat N or S 10°–20° hPa	Local time (hours)	Lat N or S 0°–10° hPa	Lat N or S 10°–20° hPa
1	-0.1	-0.1	13	0.1	0.1
2	0.3	0.3	14	0.7	0.7
3	0.7	0.7	15	1.3	1.1
4	0.8	0.8	16	1.5	1.3
5	0.6	0.6	17	1.4	1.2
6	0.2	0.2	18	1.0	0.9
7	-0.4	-0.3	19	0.5	-0.3
8	-0.9	-0.8	20	-0.1	-0.2
9	-1.3	-1.1	21	-0.6	-0.6
10	-1.4	-1.2	22	-0.9	-0.8
11	-1.1	-1.0	23	-0.9	-0.8
12	-0.6	-0.5	24	-0.6	-0.5

Table for obtaining dew point from wet and dry bulb thermometers

Dry bulb °C	Depression of the wet bulb																			Dry bulb °C
	0	0.5	1	1.5	2	2.5	3	3.5	4	4.5	5	5.5	6	6.5	7	7.5	8	8.5	9	
40	40	39	39	38	38	37	36	36	35	35	34	33	33	32	31	31	30	29	29	40
39	39	38	38	37	37	36	35	35	34	34	33	32	32	31	30	30	29	28	28	39
38	38	37	37	36	36	35	34	34	33	32	32	31	31	30	29	28	28	27	26	38
37	37	36	36	35	35	34	33	33	32	31	31	30	29	29	28	27	27	26	25	37
36	36	35	35	34	34	33	32	32	31	30	30	29	28	28	27	26	26	25	24	36
35	35	34	34	33	33	32	31	31	30	29	29	28	27	27	26	25	24	24	23	35
34	34	33	33	32	32	31	30	30	29	28	28	27	26	25	25	24	23	23	22	34
33	33	32	32	31	30	30	29	29	28	27	27	26	25	24	24	23	22	21	21	33
32	32	31	31	30	29	29	28	27	27	26	25	25	24	23	22	22	21	20	19	32
31	31	30	30	29	28	28	27	26	26	25	24	24	23	22	21	21	20	19	18	31
30	30	29	29	28	27	27	26	25	25	24	23	23	22	21	20	19	19	18	17	30
29	29	28	28	27	26	26	25	24	24	23	22	21	21	20	19	18	17	16	16	29
28	28	27	27	26	25	25	24	23	23	22	21	20	19	19	18	17	16	15	14	28
27	27	26	26	25	24	24	23	22	21	21	20	19	18	18	17	16	15	14	13	27
26	26	25	25	24	23	23	22	21	20	20	19	18	17	16	15	15	14	13	12	26
25	25	24	24	23	22	22	21	20	19	18	18	17	16	15	14	13	12	11	10	25
24	24	23	23	22	21	20	20	19	18	17	17	16	15	14	13	12	11	10	9	24
23	23	22	22	21	20	19	19	18	17	16	15	15	14	13	12	11	10	9	7	23
22	22	21	21	20	19	18	18	17	16	15	14	13	12	11	10	9	8	7	6	22
21	21	20	20	19	18	17	16	16	15	14	13	12	11	10	9	8	7	6	4	21
20	20	19	19	18	17	16	15	15	14	13	12	11	10	9	8	7	5	4	3	20
19	19	18	17	17	16	15	14	13	13	12	11	10	9	8	6	5	4	3	1	19
18	18	17	16	16	15	14	13	12	11	10	9	8	7	6	5	4	2	1	-1	18
17	17	16	15	15	14	13	12	11	10	9	8	7	6	5	4	2	1	-1	-3	17
16	16	15	14	14	13	12	11	10	9	8	7	6	5	3	2	1	-1	-3	-5	16
15	15	14	13	13	12	11	10	9	8	7	6	4	3	2	0	-1	-3	-5	-7	15
14	14	13	12	11	11	10	9	8	7	6	4	3	2	0	-1	-3	-5	-7	-9	14
13	13	12	11	10	9	9	8	6	5	4	3	2	0	-1	-3	-5	-7	-9	-12	13
12	12	11	10	9	8	7	6	5	4	3	2	0	-1	-3	-5	-7	-9	-12	-15	12
11	11	10	9	8	7	6	5	4	3	2	0	-1	-3	-5	-7	-9	-11	-15	-19	11
10	10	9	8	7	6	5	4	3	2	0	-1	-3	-5	-7	-9	-11	-14	-18		10
9	9	8	7	6	5	4	3	2	0	-1	-3	-4	-6	-8	-11	-14	-18			9
8	8	7	6	5	4	3	2	0	-1	-3	-4	-6	-8	-11	-13	-17				8
7	7	6	5	4	3	2	0	-1	-3	-4	-6	-8	-10	-13	-16					7
6	6	5	4	3	2	0	-1	-2	-4	-6	-8	-10	-13	-16						6
5	5	4	3	2	0	-1	-2	-4	-6	-7	-10	-12	-15	-19						5
4	4	3	2	1	-1	-2	-4	-5	-7	-9	-12	-14	-18							4
3	3	2	1	-1	-2	-3	-5	-7	-9	-11	-14	-17								3
2	2	1	0	-2	-3	-5	-6	-8	-10	-13	-16									2
1	1	0	-1	-3	-4	-6	-8	-10	-12	-15	-19									1
0	0	-1	-3	-4	-6	-7	-9	-12	-14	-18										0
-1	-1	-2	-4	-5	-7	-9	-11	-13	-16											-1
-2	-2	-3	-5	-6	-8	-10	-13	-15	-19											-2
-3	-3	-4	-6	-8	-10	-12	-14	-17												-3
-4	-4	-5	-7	-9	-11	-13	-16													-4
-5	-5	-7	-8	-10	-12	-15	-18													-5
-6	-6	-8	-9	-12	-14	-17														-6
-7	-7	-9	-11	-13	-15	-19														-7
-8	-8	-10	-12	-14	-17															-8
-9	-9	-11	-13	-16	-19															-9

1 Enter the temperature of the dry bulb in the extreme left or right columns.
2 Subtract wet bulb temperature from dry bulb temperature and enter on the top row.
3 Read off the dew point from the body of the table.

Conversions factors

Pressure

These days, the preferred unit of barometric pressure is the hecto Pascal (hPa), which is identical with the earlier unit, the Millibar (Mb). Some older barometers are calibrated in inches or millimetres or mercury (in Hg or mm Hg).

To convert:

> hPa to mm Hg, multiply by 0.7501
> mm Hg to hPa, multiply by 1.33316
> in Hg to hPa, multiply by 33.86

Temperature

To convert:

> Degrees Centigrade to Fahrenheit, multiply by 1.8 and add 32
> Degrees Fahrenheit to Centigrade, subtract 32 and divide by 1.8

C°	F°
0°	32°
2°	35.6°
4°	39.2°
6°	42.8°
8°	46.4°
10°	50.0°
12°	53.6°
14°	57.2°
16°	60.8°
18°	64.4°
20°	68.0°
22°	71.6°
24°	75.2°
26°	78.8°
28°	82.4°
30°	86.0°
32°	89.6°
34°	93.2°

The Beaufort Scale

Force	Description	Wind speed (kts)*	Av wave ht(m)**	Max wave ht(m)	Deep sea criteria
0	Calm	0	–	–	Sea mirror smooth
1	Light airs	1–3	0.1	0.1	Ripples but without foam or crests
2	Light breeze	4–6	0.2	0.3	Small wavelets; crests but not breaking
3	Gentle breeze	7–10	0.6	1.0	Large wavelets beginning to break, perhaps a few white horses
4	Moderate	11–16	1.0	1.5	Small waves; growing in size with frequent white horses
5	Fresh breeze	17–21	2.0	2.5	Moderate waves; taking more pronounced form with many white horses and chance of spray
6	Strong breeze	22–27	3.0	4.0	Large waves starting to form; more white foam, crests everywhere with possibility of spray
7	Near gale	28–33	4.0	5.5	Sea heaping up with white foam from breaking waves; sea begins to streak
8	Gale	34–40	5.5	7.5	Moderate light waves of great length; crest breaking into spindrift; streaking foam
9	Strong gale	41–47	7.0	10	High waves; tumbling and rolling crests, dense streaks of foam and spray affecting visibility
10	Storm	48–55	9	12.5	Very high waves with long overhanging crests; dense white streaks of foam, sea has white appearance, heavy tumbling sea, visibility affected
11	Violent storm	56–63	11	16	Exceptionally high waves; sea is completely covered with foam
12	Hurricane	64+	14	–	The air is filled with spray and visibility is seriously affected

* Measured at a height of 10 metres above sea level.
** In open sea remote from land.

Kepler elements and the NORAD element set

Johann Kepler (1571–1630) was the first person to make a thorough study of the movements of orbiting objects. He was concerned with planets and their orbits about the sun, but the methods he developed can also be applied to satellites orbiting around the earth. His discovery was that orbits can be described in terms of a set of parameters known as *Kepler elements* or *parameters*, and which still form the basis of predicting satellite motions today. Details of these are as follows:

Element	Explanation
Inclination	The angle between the satellite's orbital plane and the Earth's equatorial plane.
Eccentricity	A number between 0 and 1 describing the orbit's shape. One that is circular is zero; one that's very long and narrow is close to 1.
Right Ascension of the ascending node	The ascending node is the point on the satellite's orbit where it cuts the Earth's equatorial plane while the satellite is moving from south to north. The Right Ascension of this point is the angle formed at the centre of the Earth, between this point and the Vernal Equinox. Just as longitude is measured from the Greenwich meridian, and is one of a pair of co-ordinates used to define positions on the Earth, so Right Ascension and the Vernal Equinox are used to define positions of objects against a background of space.
Argument of Perigee	Perigee is the point on a satellite's orbit that is closest to the Earth, and apogee is a point that's furthest away. Argument of Perigee is the angle formed between a line joining this point to the Earth's centre and a line from this point to ascending node.
Mean Anomaly	This is an angle that defines the satellite's position along its orbit. Perigee is defined as 0 degrees, and apogee, 180 degrees.
Mean Motion	The number of orbital passes through the perigee per day.
Epoch	The time at which the above elements were measured.

The NORAD element set

Data for each satellite is saved in a plain text file in what is referred to as a '2 line format'. In fact, 3 lines are used to describe each satellite. The first, referred to as line 0, has a maximum of 24 characters and gives the satellite its name or identifier. Its orbital data are included in the remaining 2 lines which are 69 characters in length, as shown in the following example for a NOAA satellite:

```
NOAA   14
1 23455U 94089A   01081.91665165  .00000424  00000-0  25397-3 0  6668
2 23455  99.1725 72.3173 0009997  32.2260 327.9522 14.12609598320950
```

Data in lines 1 and 2 are organised as follows:

LINE 1

Character	Description
1	Line Number
3-7	Satellite Number
8	Classification (U = Unclassified)
10-11	International Designator (Last two digits of launch year)
12-14	International Designator (Launch number of the year)
15-17	International Designator (Piece of the launch)
19-20	Epoch Year (Last two digits of year)
21-32	Epoch (Day of the year and fractional portion of the day)
34-43	First Time Derivative of the Mean Motion
45-52	Second Time Derivative of Mean Motion (decimal point assumed)
54-61	BSTAR drag term (decimal point assumed)
63	Ephemeris type
65-68	Element number
69	Checksum (Modulo 10)
	(Letters, blanks, periods, plus signs = 0; minus signs = 1)

LINE 2

Characters	Description
1	Line Number
03-7	Satellite Number
09-16	Inclination (Degrees)
18-25	Right Ascension of the Ascending Node (Degrees)
27-33	Eccentricity (decimal point assumed)
35-42	Argument of Perigee (Degrees)
44-51	Mean Anomaly (Degrees)
53-63	Mean Motion (Revs per day)
64-68	Revolution number at epoch (Revs)
69	Checksum (Modulo 10)

Appendix 8

Internet resources

For all aspects of weatherfax use, the importance of Internet resources will no doubt continue to increase. With minimal delay before publication, it has the potential to provide the most up-to-date information on station schedules, operational status of satellites, Kepler parameters and, of course, current fax charts and space images. Unfortunately, not all Internet sites are kept up to date, and the best ones for today may not be the best ones for tomorrow. Address (URL) changes occur frequently, and as new sites emerge, the old familiar favourites sometimes disappear, often without warning. Given the temporal nature of the web, in preparing the following list I've concentrated on giving URLs of sites that seem likely to be around for some time, and that in themselves provide links to other weather-related sites and so provide access to deeper layers of information.

Understanding weatherfax
www.pangolin.co.nz/radio
Additional information on this book. Late additions and links to other sources, and software downloads including *PhysPlot* – a free code viewer.

Celestrak – Dr T S Kelso
www.celestrak.com
Satellite user support. Background documentation, Kepler elements, Frequently Asked Questions.

Marius Rentzen
www.hffax.de
A wealth of links and information on all aspects of weatherfax and satellite imagery. Includes equipment manufacturers, software downloads, fax station transmission schedules and image sources, and designs for home-built equipment.

JVComm 32
www.jvcomm.de
A popular program for reception of HF weatherfax, satellite images, and slow-scan television, reasonably priced and available in a free demonstration mode.

Airmail 2000
www.airmail2000.com
Email software used with Winlink, Sailmail and other HF email services. Included on the website is a free download of a GRIB file viewer.

Remote imaging group (RIG)
www.rig.org.uk
A satellite user support group providing a wealth of information and links to other sources.

National Weather Service Home Page
www.nws.noaa.gov/
The US National Weather Service site is enormous. It includes not only US weather, but much of international interest and links to other meteorological sites around the world. Use the following URLs to cover marine and weatherfax services.
www.nws.noaa.gov/om/marine/radiofax.htm
www.nws.noaa.gov/om/marine/home.htm

Appendix 9

Bibliography

Weather Maps – How to Read and Interpret all the basic Weather Charts, Chaston Scientific Inc, Peter R Chaston, PO Box 758, Kearney MO 64060, USA.
An Introduction to Satellite Image Interpretation, Eric D Conway and the Maryland Space Grant Consortium, Johns Hopkins University Press, Baltimore and London, 1997.
MetService Yacht Pack, Bob McDavitt, Capt Teach Publications.
Images in Weather Forecasting – A Practical Guide for Interpreting Satellite and Radar Imagery, M J Bader, G S Forbes, J R Grant, R B E Lilley and A J Waters.
Weather Predicting Simplified; How to Read Weather Charts and Satellite Images, Michael William Carr, International Marine/McGraw-Hill, 1999 (ISBN 0-07-012034-5).
NSW Coroner's report into the 1998 Sydney–Hobart Race, John Abernethy, NSW State Coroner, December 2000.
The Science and Wonders of the Atmosphere, S D Gedzelman, John Wiley & Sons, Chichester, 1980.

Index